Culture and Sustainability

Janet Stephenson

Culture and Sustainability

Exploring Stability and Transformation with the Cultures Framework

palgrave
macmillan

Janet Stephenson
Centre for Sustainability
University of Otago
Dunedin, New Zealand

ISBN 978-3-031-25514-4 ISBN 978-3-031-25515-1 (eBook)
https://doi.org/10.1007/978-3-031-25515-1

Cover credit: Jenna Packer, Bread and Circuses (2013/14), acrylic on canvas, 1205 × 1601 × 45 mm. Private Collection. Courtesy of Milford Galleries Dunedin. Photograph: Glenn Frei

This Palgrave Macmillan imprint is published by the registered company Springer Nature Switzerland AG
The registered company address is: Gewerbestrasse 11, 6330 Cham, Switzerland

Preface and Acknowledgments

This book might have been different had it been written it from another place in the world. Here in Aotearoa New Zealand, a small island nation in the south-west Pacific Ocean, there is a sense of being at the edge: less constrained by Western academic norms and freer to experiment with ideas. Here, Māori knowledge systems and worldviews increasingly permeate academia, challenging traditional thinking and inspiring new approaches. Sustainability issues are to the fore in politics and policies. Our physical isolation can mean that time- and place-specific academic preoccupations are felt less intensely. Perhaps this is why my book is less concerned with disciplinary boundaries and theoretical disputes, and more focused on drawing insights from across disciplines and from non-Western knowledge systems. But despite being written in this place, it is not a book about this place. *Culture and Sustainability* is universally relevant. It takes an inclusive approach to culture—in all its senses and academic renditions—and a broad perspective on sustainability.

The seriousness of the global sustainability crisis cannot be underestimated. Actions over the next 10–15 years will determine whether we deliver a resource-depleted, over-heated and destabilised planet to coming generations or whether we can claw our way to a sustainable future. My primary motivation in writing this book was to amplify the role of culture at this crux point. Culture is simultaneously a cause of the crisis, a constraint on change, an exemplar of solutions and a creative force for transition. Yet culture's causal role and transformational potential are largely unappreciated beyond a scattering of academics. This book aims to make the concept of culture more accessible and applicable as an analytical frame which anyone can use to do work for the sustainability transition.

Culture and Sustainability has two broad themes. One theme explores the many lay and academic interpretations of culture. Here I attempt to make sense of culture for non-specialist readers and to show how all approaches to culture can potentially contribute to the daunting task of sustainability. The other theme elaborates on the cultures framework: a particular approach to analysing sustainability problems using a cultural lens. The cultures framework began life as the energy cultures framework and was first introduced in academic literature in 2010. It was subsequently adopted by numerous researchers to explore the role of culture in sustainability-related issues in both developed and developing countries. I am grateful to and inspired by the international research community for this 'road testing' of the framework, the results of which can be seen in the many case studies which enliven the book.

The foundational work of developing the framework was undertaken by the Energy Cultures research teams between 2008 and 2016, where we developed and trialled a novel approach to investigating energy culture and applied it to energy and transport issues. I led and co-led an extraordinary team whose disciplinary backgrounds included sociology, economics, physics, psychology, engineering, law, marketing, management, system dynamics and human geography. We learned a huge amount from each other, had a lot of fun and discovered that the cultures framework was as relevant to our own teamwork as it was to the problems to which it was being applied. This book pays homage to the friendships, support and intellectual contributions of the Energy Cultures teams: Rob Lawson, Gerry Carrington, Barry Barton, Paul Thorsnes, Miranda Mirosa, Debbie Hopkins, Martha Bell, Sara Walton, John Williams, Rebecca Ford, David Rees, Alaric McCarthy, Mary Jo Lavelle, Michelle Scott, Ben Wooliscroft, Sally Blackwell, Adam Doering, Gilles Marciniak, Ikerne Aguirre-Bielschowsky, Seth Gorrie, Geoff King, Sam Spector and Jane Khan.

My more recent collaborations have applied the cultures framework in new and fruitful ways, and for this my sincere thanks go to Benjamin Sovacool, Tor Håkon Jackson Inderberg, Gerry Carrington, Debbie Hopkins, Linda Bach, Michael Jack, Ben Anderson, Hugh Campbell, Abbi Virens and Scott Willis.

I have also been inspired by the work of my postgraduate students who have used the cultures framework in their own work: Ph.D. students Daniel Gnoth, Imran Khan, Will Stovall, Carsten Dortans, Kakau Foliaki and Jefferson Dew, and Masters students Letisha Nicholas and Diana Giraldo Ocampo. Many other postgraduates who have used the framework have reached out to me from different parts of the world to ask questions or share their findings. All have expanded my thinking through their diverse applications of the framework and our many discussions.

One strand of the book draws on the paradigm-breaking contributions of Indigenous knowledge. I am indebted to everyone who has contributed to my awareness in this respect, including research collaborators and co-authors Rauru Kirikiri, Jacinta Ruru, Merata Kawharu, Lyn Carter, Jonathan Dick, Corey Bragg, Stephanie Rotarangi, Haerewa Ngarangi, Henrik Moller, Rachel Turner, Fikret Berkes, Nancy Turner, Kyle Artelle and all of the Project Kainga team. Ngā mihi nui ki a koutou.

The development and testing of the energy cultures framework was supported by several grants for which I am most grateful. The first field-testing of the concept was supported by a small University of Otago Research Grant in 2008. Soon after, we received significant national funding for an applied research programme (Energy Cultures 1) from New Zealand's Foundation for Research, Science and Technology over 2009–2012. A further major grant from the Ministry of Business, Innovation and Employment funded the Energy Cultures 2 research programme over 2012–2016. Financial support from the Energy Efficiency and Conservation Authority and the Ministry of Transport enabled some associated projects. I also acknowledge the University of Otago's support for a year-long sabbatical which gave me the time and intellectual space to make significant progress with this book.

To my wonderful colleagues at the Centre for Sustainability, thank you for your interest and support as I explored ideas and worked on the manuscript. My sincere thanks to Will Stovall for his invaluable work as a research assistant for this book and for his helpful comments on a late draft. Special thanks to Jenna Packer who kindly agreed to my use of her evocative painting for the front cover. To the reviewers of the draft manuscript, thank you so much for your pertinent and valuable feedback. To my friends and family, thank you for putting up with me when I was deep in thought or tentatively trying out new ideas on you, and for distracting me when I needed it. And for my husband Guy, who plied me with endless cups of tea and unwavering support, my deep and abiding love.

Dunedin, New Zealand Janet Stephenson

Contents

List of Figures

List of Tables

CHAPTER 1

Introduction

A Crisis of Culture

The future looks bleak. Regardless of where we are in the world, we face a changing climate, depleted resources, widespread pollution, the exacerbation of inequities and consequential challenges to health, welfare and geopolitical stability. Humanity is facing a sustainability crisis. I believe that this is largely a crisis of culture.

Culture is complicit because the origins of our sustainability predicament lie in the cultural enactment of ideologies of over-consumption, capitalism and colonialism. Cultural processes can constrain transformational change because of the tenacity of deeply embedded patterns of beliefs, practices and material expectations. Yet culture can also be a powerful force for change. Culture can quickly adapt to new circumstances, and much can be learned from the many cultures that have long known what it takes to live sustainably.

Reader, I feel you nodding. Such sweeping statements are easy to agree with. But what does culture mean in these contexts? How do we even understand culture?

In a lay sense, culture is an attractive concept when it comes to explaining sustainability problems. It is a convenient catch-all term for everything that is mystifying or hard to change about human collective behaviour. It can be used to generalise the problem to a point of abstraction (e.g. 'the problem is Western culture') or to over-simplify the solution (e.g. 'we just have to change our culture'). It is used as a device

J. Stephenson, *Culture and Sustainability*,
https://doi.org/10.1007/978-3-031-25515-1_1

for directing criticism at others (e.g. 'their organisational culture is terrible') or for deflecting criticism away from ourselves (e.g. 'that's just our culture'). It is often used tautologically ('culture is the problem but that's the way culture works') or as a reason for inaction ('it's culture so it's not going to change'). It is afflicted with the curse of common sense ('of course everyone knows what culture is') and the curse of the residual ('it's everything about humans that we can't explain or don't understand'). So while culture is an attractive idea, using it simplistically can be worse than useless: it actively supports the unsustainable status quo by giving it an excuse.

If culture is an elusive concept in its lay sense, it is many times worse in its academic sense. Culture is claimed and used in very specific ways in different disciplines, including anthropology, sociology, cultural studies, media studies, business management, consumption studies, creative arts and in the biological sciences. Some disciplines have developed such specialised terminologies to describe their interpretation of culture that their work has essentially become a closed shop, inaccessible to those in other disciplines. Within disciplines there may be further divergences of meaning based on obscure points of theory. To compound its slipperiness, I have noticed that culture is often undefined in academic articles, books and policy reports, leaving it to the reader to bring their own assumptions and understandings. It is hardly surprising that academic and cultural critic Raymond Williams famously called culture 'one of the two or three most complicated words in the English language' (Williams, 1976: 76).

Culture deserves to be better understood. Like a deep, slow river, it has a deceptively smooth surface that reflects back what we bring to it. For culture to become a useful concept for sustainability analysis, we need to peer into the depths to become aware of culture's many channels, divergences and undercurrents. We need to look upstream to become aware of its origins and its journey over time and space. We need to find ways of harnessing its power to help navigate our way to a sustainable future.

This book is concerned not just with culture's meaning, but also with what culture is and does. At its most fundamental, culture is about difference between groups of people. We are aware that our partner's family have their own rituals around meals and celebrations. We notice when people wear clothes or eat food that are beyond our concept of normal. We can see that our friend's workplace is less hierarchical than ours. Some of the words and symbols used by teenagers are a mystery to us. And yet our own way of life seems entirely normal. It can be hard to see our own culture without stepping outside of it.

This very quality of differentiation means that culture can be an ideological weapon. The colonial project was in large part driven by the conviction that Western culture was the peak of civilisation and Western nations therefore had the right to global domination. The resulting cultural hegemony led to the destruction and ongoing domination and repression of many non-Western societies. Cultural difference is at the heart of othering: in beliefs about entitlement, in cultural appropriation and in culture wars. Culture is intimately related to power.

If we look more deeply again, culture reflects one side of the foundational nature–culture schism in dominant Western conceptualisations of the world. Nature is perceived to be fundamentally separate from humans and their ideas, beliefs, actions and products. This nature–culture dichotomy shaped Western science, impelled particular forms of economy and business, and underpins the belief that nature is infinitely replaceable by human ingenuity: a belief whose danger we are now learning to our peril. This duality shapes most European language systems, such that word couplings like 'bio-cultural' or 'social-ecological' are required to refer to the wholeness of existence. Through the language used by many of us, this binary code creates habits of thought that are hard to elude. In contrast, many Indigenous societies perceive people and nature to be seamlessly entangled, and this is reflected in their languages and practices. Within many Indigenous worldviews, there is no directly equivalent term to culture that sets it as separate to nature, because elements of nature are kin. Those of us accustomed to a Western worldview must take care that we are not automatically dismissing non-Western ontologies in our conceptualisations of culture.

So while it may be easy to agree that culture is a critical factor in transitioning to a sustainable future, it is not obvious what to do next. Used in a lay sense, culture can be invisible, unhelpful and unsafe. In the scholarly realm, culture is fragmented by academic understandings and often couched in inaccessible language. Where it is used, it is often undefined which increases the risk of scholars talking past each other. Cultures are diverse, and the identification of difference brings the risk of discrimination and domination. In addition, culture is preloaded with a binary perspective of a world, which is increasingly untenable in addressing the sustainability crisis.

Culture and Sustainability seeks to address this impasse. It aims to make culture an accessible and useable concept for researchers and practitioners working on sustainability challenges. To assist with this, the book describes a framework for cultural analysis that was initially developed to study sustainability issues in energy and transport in New Zealand (Stephenson, 2018, 2020; Stephenson et al., 2010, 2015) and now is widely used in research on sustainability questions and the development of policy advice. The framework has proved particularly fruitful in analysing culture and its relationship to sustainability outcomes, regardless of topic, scale or location. *Culture and Sustainability* draws from this extensive body of work, with three main aims: to link the cultures framework more closely to the multiverse of scholarly literature on culture; to further develop and expound the framework as a basis for research and as a scaffolding for interdisciplinary and multi-theoretical studies; and to show how the framework can be used to underpin research and policy interventions to support sustainability transitions.

I wrote this book because I am deeply concerned about humanity's unsustainable trajectory and our apparent inability to change course rapidly enough to avoid catastrophe. I believe that those who have capacity to act to redirect this trajectory have a responsibility to do so. In particular, those of us who are privileged to be in academic or professional roles bear a particular obligation as conscious and informed actors who have personally benefitted from unsustainable systems over our lifetimes.

The book was written with an audience of four in mind: a postgraduate student, a senior policy advisor, the leader of an interdisciplinary research project and an academic specialising in cultural theory. My goal was to offer the student a foundation for their research methodology; to help the practitioner to design policy for sustainability interventions; to show the research leader how the framework can support a multi-method, integrated research programme; and to offer sufficient theoretical robustness and novelty to pique the cultural academic's interest. In addition, I wanted the book to be accessible to lay readers interested in culture and/or sustainability challenges. Where I refer to 'we' in the book, I mean myself and all my readers.

Over the rest of this chapter, I will revisit why cultural analysis has such a key role in achieving a sustainable future and briefly introduce my own work with interdisciplinary research teams that led to the development of the cultures framework. But first, I return to the meanings of culture.

What Is Culture?

Having thoroughly confused my readers with culture's complexity and slipperiness, it is only fair that I come to your rescue and offer (at least for the time being) a foundation on which to start. In the social sciences, and particularly in anthropology and sociology, culture is a core explanatory concept. It usually refers to shared qualities and social processes that are neither unique to individuals nor common to humanity as a whole. Culture is about how social groups develop and retain distinctive features; how they convey meanings and identity; and how they maintain membership over time. Cultural processes underpin the differences across human societies as well as the relative homogeneity within any given group of social actors.

Put very simply, culture comprises similar patterns across a group of people in how they think, what they do and what they have. In later chapters, I will complicate and refine this notion, but it is a good place to start. Culture reflects the fact that what we think, do and have (at home, in the workplace, online, in the community, with our friends) is strongly influenced by others. Culture comprises ways of understanding and thinking about the world that we share with others and which motivate us to act as we do, actions that we undertake on a regular or irregular basis, and things that we make, use, appreciate or acquire. An important feature of culture is how these are interconnected: shared cognitive features shape what we do and what we have; actions we learn from others shape how we think and what we have; and what we have shapes what we do and how we think. These dynamics are indicated in Fig. 1.1.

Exploring the implications of culture for sustainability is a somewhat different task to examining culture more generally. Rather than examining cultures that are definable by, say, particular musical tastes or beliefs about the afterlife, I am interested in aspects of culture(s) that have implications for societal, environmental and/or economic wellbeing. There is a sturdy thread of such studies, such as work on cultures of consumption (McCracken, 1990), cultures of waste (Hawkins & Muecke, 2003), petrocultures (Wilson et al., 2017) and on automobility, the cultural norming of car ownership (Urry, 2004). Culture also has positive associations with a more sustainable future, such as the worldviews, values and associated practices of at least some Indigenous cultures (Artelle et al., 2018; Berkes, 1999) and those of groups who work on environmental causes (e.g. Peace et al., 2012) or intentionally change their behaviours

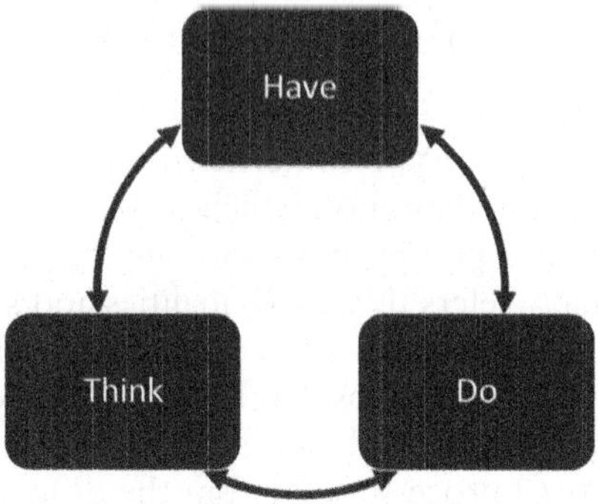

Fig. 1.1 Fundamental elements and dynamics of culture

and consumption choices (e.g. Quinn & Westwood, 2018). At a governance scale, actions and decisions are shaped by 'political cultures' (Geels et al., 2007) and 'policy cultures' (Bailey, 2007), and nations may have different 'energy cultures' (Stephenson et al., 2021).

In all these senses and more, culture can be seen to play an important role in causing the unsustainable world we now inhabit and to have the potential to play an even more important role in transitioning to a sustainable one (Goggins et al., 2022; Sovacool & Griffiths, 2020a, 2020b). The past ten years have seen a surge of interest in social science research on culture, much of it in the field of energy and transport. However, culture is largely overlooked in most sustainability literature or is placed in a residual category of influences rather than being recognised as the powerful force that it is. Accordingly, this book aims to make culture more accessible as a concept, and more effective as an analytical lens.

Why Culture Matters

The nature of the sustainability crisis is well traversed elsewhere and does not need to be repeated here in any detail: it is a cascading destabilisation of Earth's natural systems, with terrifying implications for the health and wellbeing of humans and other living things. The crisis tends to be described through partial descriptors such as climate change, mass extinctions, pollution, deforestation or ocean acidification, and its human impacts through labels like food crises, climate migration, financial crashes, geopolitical tensions and intergenerational inequities (United

Nations, 2015). The crisis, however, does not come in neat packages. It comprises the totality of these and other measures of destabilisation and the linkages between them (Lade et al., 2020; Steffen et al., 2015). Its fundamental causes are systems of exploitation, production and consumption that are founded in a belief that Earth's systems have an infinite capacity to act as a source of resources and a sink for waste (Klein, 2015).

Responsibility for the sustainability crisis is unevenly shared. Considering just greenhouse gas emissions, the world's wealthiest countries (including the United States, Canada, Japan and much of western Europe) are responsible for 50% of all greenhouse gases (GHG) produced from industry and fossil fuels since 1850, despite comprising only 12% of the global population (Andrew & Peters, 2021). Analysed in relation to businesses, two-thirds of historical greenhouse gas emissions between 1880 to 2010 were produced by only 90 companies (Ekwurzel et al., 2017), and since 1998, 71% of the global GHG emissions have been produced by just 100 companies (Griffin, 2017). Looking at comparative responsibility another way, the wealthiest 10% of the world's population were responsible for around 50% of global emissions in 2020, with the wealthiest 1% responsible for 15% of emissions. In contrast, the world's poorest 50% were responsible for only 7% of global GHG emissions (Kartha et al., 2020).

Accordingly, a small proportion of nations, corporations and individuals are the prime promulgators (and beneficiaries) of unsustainable systems of production and consumption. Their motivations, actions and material choices have dominated the degradation of the world's natural systems as well as amassing wealth in few hands. Capitalism inherently requires inequalities in order to thrive (Piketty, 2020). Because of these actors and their actions, much of the remainder of humanity has either been exploited and distanced from any ability to establish a modest sustainable livelihood, or has become inexorably captured by the pursuit of unsustainable consumption, such that living beyond planetary limits has become normalised and largely unquestionable. Despite widespread knowledge of the damage they are causing, these powerful actors have continued to enrich themselves and exploit others at the cost of the stability of global systems (Dunlap & Brulle, 2020). This is implicitly supported by the widely shared belief that humans and organisations have the right to pursue their wealth and power objectives despite the extreme costs borne by the rest of humanity and other species, now and into the future.

It is still possible to change direction, but, as is evident from the recent reports of the Intergovernmental Panel on Climate Change and the United Nations, we have very little time left (IPCC, 2022; United Nations, 2022). Within one generation (25 years), the world needs to achieve net zero carbon emissions alongside markedly reducing other critical exceedances. This will require radical changes to established ways of sourcing materials, providing services and creating products as well as to accepted levels of consumption by the (relatively) wealthy. It will require transformations in all sectors of society including governance agencies, businesses, communities and households. The scope of the challenge cannot be underestimated. It involves restoring the integrity of environmental systems while grappling with the impacts of climate change and other self-inflicted destabilisations, along with developing sustainable livelihoods and resilient communities, and ensuring that inequities are addressed or at least not exacerbated. If these radical changes are not achieved, more radical changes will be forced upon us all by an unravelling planet.

Visions of how deep-seated the transition needs to be are hugely varied. The most widely promoted concept of a sustainable future is captured by the Sustainable Development Goals (SDGs). These were adopted by all United Nations Member States as part of the 2030 Agenda for Sustainable Development (United Nations, 2015). The seventeen SDGs include ending poverty and other deprivations, improving health and education, reducing inequality, tackling climate change and preserving oceans and forests, while at the same time spurring economic growth. A similar approach is taken by the International Energy Agency, which aligns renewable energy expansion with jobs and economic growth (International Energy Agency, 2020). In these visions, sustainability means the concurrent pursuit of social, economic and environmental goals, while overlooking that many will be incommensurable and thus require trade-offs (Spangenberg, 2017). If the main measure of success is economic growth, then it is inevitable that social and environmental goals will suffer.

Yet it is increasingly clear that aspirations for everlasting economic growth as promoted by orthodox economics will need to be tempered by biophysical realities (Boston, 2022). An alternative perspective recognises that the functions of natural systems cannot be substituted, that growth and resource use cannot be sufficiently decoupled, and that humans must constrain their consumption to avoid destroying the natural processes on which we depend (Ayres et al., 2001; Parrique et al., 2019).

This is reflected in 'doughnut economics' which takes the position that economic activity should operate within a safe and just space for humanity. Consumption aspirations must be tempered so that all people have access to the essentials of life (e.g. sufficient food, housing, a political voice, adequate health care) while not overshooting Earth's natural systems (Raworth, 2017).

Other analysts call for an even more profound re-visioning of local and global economies, arguing that to live within biophysical limits will involve radical reductions in expectations of financial wealth and consumer items, and living in simpler, more locally focused ways. This will require degrowth (i.e. purposeful reductions in GDP) and is argued by some to require new economic systems such as post-growth or post-capitalist economies (Büchs & Koch, 2019; Jackson, 2021; Kallis, 2018).

Regardless of how far-reaching the changes will need to be, the process of getting there will be complicated and hard. It will involve the transformation of systems of production and consumption that have caused the global sustainability crisis. It requires major adjustments away from business-as-usual at all scales, and the realisation of ideas that only a decade ago seemed radical, such as embargoes on new fossil fuel developments, circular economies, net zero housing, carbon-negative cities and regenerative agriculture. Most fundamentally, it requires discarding ideologies that have underpinned the unsustainable economic system over the past 200 years, and developing new shared ideologies, institutions and practices that will enable humans to live within planetary limits and ensure that sustainability transitions incorporate and achieve distributional, procedural and restorative justice (McCauley & Heffron, 2018).

Unfortunately, there is insufficient traction with any of these ambitions. The United Nations reports that the sustainable development goals for 2030 seem increasingly unachievable, and that humanity's own survival is in grave danger as a result of interlinked and cascading crises (United Nations, 2022). The rate of species extinctions is accelerating and is hundreds of times higher than in the past 10 million years (IPBES, 2019). If current country policies continue, the globe is on track for a temperature increase of between 2 and 3.6 degrees by 2100. Even if pledges and targets made as part of countries' nationally determined contributions are put into effect, there will still be a rise of around 2.1 degrees by the end of the century (Climate Analytics & New Climate Institute, 2022). Despite their apparent intent to take action, nations and sectors are not changing quickly enough and are unwilling to undertake change at the scale and depth required.

We can point the finger of blame in many directions: at politicians who fear backlash from their voters; at powerful industries that see most profits in the status quo; at economic systems that favour short-term profit over long-term benefits; at business sectors that keep doing the same thing because it is the easiest; at policies that give favour to business-as-usual; at social media for promulgating denial and unrealistic expectations; or at householders who aspire to live what, to them, seems to be a normal life. While some are more responsible than others, all are complicit in perpetuating the crisis, and none can solely deliver the solutions we need. Across all of these actors, the fundamental issue is the intractability of unsustainable beliefs, ideologies, values, practices and material expectations—a crisis of culture.

Culture is fundamental to the transformative changes that are needed to avert the worst impacts of a destabilising planet and to create a more hopeful and just future. Sustainability transitions, as they are often called, necessarily involve societal transformations as well as political, economic and technological ones. Most perspectives on transitions to date have paid little attention to culture, although some work in the field of socio-technical transitions has explored aspects of culture's influence within broader processes (e.g. Geels & Verhees, 2011; Sovacool & Griffiths, 2020a, 2020b). Winkler (2020) describes the current crisis as primarily resulting from cultural hegemony by the ruling class who share an ideology that centralises economic growth, with unsustainable consequences that include a damaged environment and social inequities. He proposes that this dominant culture can only be altered by change agents from civil society, business and government who promote the destabilisation of the dominant regime and its replacement with new ideologies and related material and non-material conditions that align with sustainable outcomes. Purposeful cultural change at multiple scales is critical to achieving sustainability transitions.

Introducing the Cultures Framework

Culture—as it relates to sustainability—urgently needs to be better understood. If we are to have a chance of getting to the end of this century without completely undermining planetary life-support systems, we need to understand how culture operates. We need to know how and why some cultures change while others remain relatively static. We need to discover how more sustainable cultures have emerged in the past, and whether and how they have maintained these characteristics. We need to understand the reach and influence of powerful unsustainable cultures, and how these might be refashioned. We need to see how new cultural arrangements—new ways of thinking, doing and having—are sparked, emerge and spread.

But we have a problem. Culture is one of the most widely used concepts in the social sciences, yet academics disagree on what it means, and lay people have vague and varied understandings. Academics can be critical of how non-specialists use the concept of culture, while, from a lay perspective, academic preciousness and jargon make its meaning almost impossible to penetrate. How, then, to effectively use the concept of culture as an analytical framing to assist in achieving the sustainability transition? Instead of being obscure and unreachable, or alternatively used as an excuse or an accusation, how can we make culture comprehensible and applicable by anyone to the sustainability problems they face?

The cultures framework (Fig. 1.2) is an extensively tested answer to these questions. It is introduced fully in Chapter 4 and elaborated with examples in Chapters 5–8. It is a conceptual framework representing the dynamic ensemble of culture and its relationship with sustainability outcomes. As shown in Fig. 1.2, it draws attention to broader influences on culture that may act to reinforce cultural patterns or alternatively may enable change. The framework centres on any actors in whom we are interested—such as individuals, households, communities, organisations, businesses or governments—and takes account of their agency (their ability to make choices and act) because ultimately the sustainability transition will not occur without actors having the ability to make purposeful change to their cultural ensembles.

The cultures framework is simultaneously a framework to support interdisciplinary, multi-theory investigations into culture, and a model that can be used for analysis in its own right. At the time of writing, it had been used to underpin or inform over 100 research projects in more than 30 countries by researchers from a wide range of knowledge systems and disciplines. It has been shown to provide fruitful insights into both cultural inertia and processes of cultural change. At its most basic, the framework helps researchers and lay people to 'see' culture as a

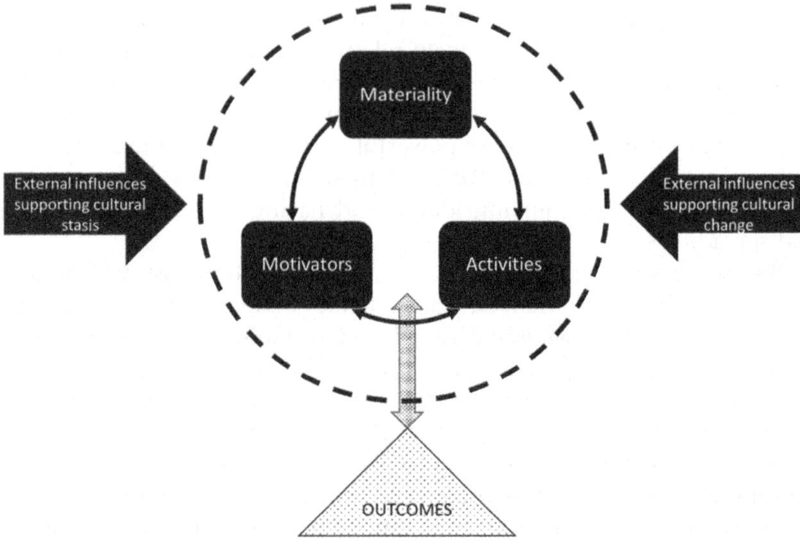

Fig. 1.2 The cultures framework

tangible, dynamic process rather than a puzzling and immutable feature of the social realm.

I played a leading role in the development, testing and refinement of what we initially called the 'energy cultures framework' in the interdisciplinary Energy Cultures research programmes which we carried out in New Zealand between 2008 and 2016. Our core interest at that time was in exploring why it was so difficult for households and businesses to change to more energy-efficient behaviours and/or technologies. Existing discipline-based analytical approaches offered only partial insights into why change was so hard. We looked at explanations from microeconomics, behavioural economics, technology adoption models, social and environmental psychology and various sociological theories including socio-technical systems and practice theories. Each offered a window on an aspect of behaviour change, but most did not deal well with the fact that humans live within a complex system of influences that cannot be reduced to simple linear explanations of cause and effect. We had a hunch that the messy interplay of personal, societal and technological factors—including social norms, people's relationships with technologies, organisational behaviours and broad-scale ideologies and institutions—was strongly influential in maintaining the status quo. But while aspects of this puzzling mélange were well examined through particular disciplines or theories, we lacked a conceptual framework to consider this complex of influences as a whole.

From its first iteration, the cultures framework drew from multiple theories and explanations of behaviour from the social sciences, and aimed to span research traditions centred on the individual and those focused on wider social processes (detailed in Stephenson, 2018; Stephenson et al., 2010; Stephenson, Barton, et al., 2015). Its theoretical roots included concepts of habitus (Bourdieu, 1977), structuration (Giddens, 1984), practice (Reckwitz, 2002; Shove & Spurling, 2013), lifestyles (Chaney, 1996), socio-technical systems (Geels, 2002), actor-network theory (Latour, 1996), systems thinking (Midgely, 2003) and anthropological and sociological approaches to culture (Hays, 1994; Ortner, 1984). From the outset, the framework depicted culture as the interplay between how people think and what they do and have. The terminology of this triad has evolved as the framework has been tested, revised and applied to an increasingly wider range of topics, but the concept of three core elements of culture has endured.

Team members in the first major Energy Cultures research programme were from the disciplines of physics, economics, law, consumer psychology and sociology. The cultures framework not only offered a conceptual centrepiece for the design of a multidisciplinary research programme, but also supported interdisciplinary collaborations, the integration of findings and the development of policy advice. The second Energy Cultures research programme involved an even broader array of disciplines and covered questions to do with efficiency, sustainability and technology adoption in the energy and transport sectors, and focused on both households and businesses as cultural actors. The cultures framework continued to prove its value as a conceptual framing and analytical model, and as an integrator of findings and the basis of further policy briefs (Stephenson, 2018, 2020).

Over the seven years of these two programmes, the framework underpinned research findings on many different topics including household energy efficiency (Lawson & Williams, 2012), solar lighting (Walton et al., 2014), timber drying (Bell et al., 2014), youth mobility (Hopkins & Stephenson, 2014, 2016), urban freight (Hopkins &

McCarthy, 2016), household solar generation (Ford et al., 2017), transport transitions (Stephenson, Hopkins, et al., 2015), household energy efficiency interventions (Scott et al., 2016), driver efficiency (Scott & Lawson, 2018), business energy efficiency (Walton et al., 2020) and policy advice (Stephenson et al., 2016). These applications tested its 'completeness' as a model as well as its ability to produce reliable and useful findings. Importantly, we started to realise how well it communicated the concept of culture (and the implications of cultural dynamics) to policymakers, politicians and community members. They were easily able to pick it up as a thinking tool and see its relevance to their fields of interest.

Since the end of the research programme, I and some of the other original team members have continued to use the framework in our own work, as have some of our postgraduate students. But excitingly, the framework took on a life of its own. From an early stage, it was adopted by other researchers internationally, often applied to energy-related topics but also to issues as varied as personal reduction of greenhouse gas emissions (Young & Middlemiss, 2012), domestic water demand (Manouseli et al., 2018), cooking (Jürisoo et al., 2019) and gender (Johnson et al., 2019). It was clear that the framework produced useful and insightful findings across a far wider range of fields than we had envisaged. By 2018 I was writing about it as a generic framework that could help investigate 'sustainability cultures', as most of its applications were in the general field of sustainability (Stephenson, 2018, 2020).

The framework consistently offers insights into how culture operates to constrain and/or enable human and organisational change. It has been applied at scales from the intimate (e.g. the lives of families in fuel poverty [McKague et al., 2016]) to pan-national comparisons of decarbonisation pathways (Stephenson et al., 2021). It has been used by researchers from a broad range of disciplines, and frequently by interdisciplinary teams. As I will elaborate in Chapter 8, most studies have used qualitative methods, either designing interviews around the elements of the framework (e.g. Hopkins & McCarthy, 2016) or applying the framework retrospectively to analyse qualitative data (e.g. Dew et al., 2017). Others have used quantitative methods, for example using national demographic data sets to identify clusters of actors with similar cultural characteristics (e.g. Bardazzi & Pazienza, 2017) or using the framework to integrate large quantitative data sets (e.g. Manouseli et al., 2018). Still others have used a mix of qualitative and quantitative methods (e.g. Scott et al., 2016). Findings are consistently fruitful and provide insights not achievable through other methods of analysis (Dew et al., 2017).

Writing this book has provided an opportunity to review this large body of work that has used, tested and extended the cultures framework. I am deeply grateful to the many authors whose work I have mined for their insights and suggestions. Writing the book over a period of sabbatical also gave me the luxury of time to return to the origins and evolution of culture as a concept. These two sources have been incredibly rewarding because they enabled me to take a fresh look at the framework to see how it stacked up, both in a theoretical sense (relating to theories of culture) and in a practical sense (relating to the insights from its many users). In Chapter 4, I present the outcomes of this exercise, revealing some additions and adjustments to earlier iterations of the framework, but with its core set of concepts having survived the test.

Conclusion

We need to better understand the role that culture plays in our unsustainable past and present, and the part it will play in achieving a sustainable future. Regardless of the scope and scale of the transition that we face—and this will vary from country to country and community to community—culture will be central to failure and at the core of success. We will not achieve sustainability transitions if we fail to understand how culture operates: as a force of power and dominance, and as a source of wisdom and inspiration; as a reproducer of the status quo, and as a creative force for transformation.

Using culture as an overarching frame is particularly advantageous for sustainability inquiries, as it enables a coherent interpretation of a complex array of influences on a given outcome. The cultures framework has shown promise in this regard, but so too do other interpretations of culture. The next two chapters focus on the concept of culture and trace its etymology and academic evolution into multiple streams of meaning. Chapter 2 discusses divergence and differences, and Chapter 3 explores common threads across these meanings. In this sense, the two chapters work as a tapestry—Chapter 2 as the warp and Chapter 3 the weft. Out of this exercise, certain patterns appear which find their way into Chapter 4.

In Chapter 4, I fully introduce the cultures framework as an analytical frame for the study of culture in relation to sustainability. I start by discussing the difference between theories and frameworks, and the concept of causality in social research. I then proceed to describe each element of the framework in some detail, as well as how the framework operates as a whole.

Chapters 5 and 6 are full of stories. They draw mainly from the many applications of the cultures framework in research projects around the world, describing how it has been used and the understandings it has delivered. I also draw out more general insights from across the studies about how culture operates. Chapter 5 covers cultural stability—how and why cultures tend to retain the same or similar patterns of features over time and tend to resist change. Chapter 6 explores cultural change through examples of minor or major cultural transformations that have improved sustainability outcomes, and the processes by which this has occurred.

The cultures framework has proved particularly helpful in identifying opportunities for policy design or other actions to support cultural transformations. Drawing from research-based examples, Chapter 7 discusses how to use the framework for policy design and evaluation, as well as more generally to assist groups and organisations to notice their own culture and potential avenues for change.

In Chapter 8, I discuss how the framework can be used to underpin research, and the kinds of questions it can be used to explore. I show how it can be used as a model by a single researcher or as an integrating framing for interdisciplinary (multi-theoretical) research. Using examples, I discuss the range of qualitative and quantitative methods that have been effectively used with the framework to date. I suggest ways in which the various interpretations of culture introduced in Chapter 2 can contribute to the sustainability research agenda.

Chapter 9 draws the book to a close. I reflect on culture through a sustainability lens and sustainability through a cultural lens. I discuss the implications of cultural processes for local and global sustainability challenges. I suggest further potential applications of the framework and further testing of its propositions. Although I am excited at the potential of the cultures framework to underpin sound research, policy and action, I believe that all forms of cultural scholarship have much to contribute to resolving the sustainability crisis.

References

Andrew, R. M., & Peters, G. P. (2021). *The Global Carbon Project's fossil CO_2 emissions dataset (2021v34)*. https://doi.org/10.5281/zenodo.5569235

Artelle, K. A., Stephenson, J., Bragg, C., Housty, J. A., Housty, W. G., Kawharu, M., & Turner, N. J. (2018). Values-led management: The guidance of place-based values in environmental relationships of the past, present, and future. *Ecology and Society, 23*(3), 35.

Ayres, R., Van den Berrgh, J., & Gowdy, J. (2001). Strong versus weak sustainability: Economics, natural sciences, and consilience. *Environmental Ethics, 23*(2), 155–168.

Bailey, I. (2007). Market environmentalism, new environmental policy instruments, and climate policy in the United Kingdom and Germany. *Annals of the Association of American Geographers, 97*(3), 530–550.

Bardazzi, R., & Pazienza, M. G. (2017). Switch off the light, please! Energy use, aging population and consumption habits. *Energy Economics, 65*, 161–171.

Bell, M., Carrington, G., Lawson, R., & Stephenson, J. (2014). Socio-technical barriers to the use of energy-efficient timber drying technology in New Zealand. *Energy Policy, 67*, 747–755.

Berkes, F. (1999). *Sacred ecology: Traditional ecological knowledge and resource management*. Taylor & Francis.

Boston, J. (2022). Living within biophysical limits: Green growth versus degrowth. *Policy Quarterly, 18*(2), 81–92.

Bourdieu, P. (1977). *Outline of a theory of practice*. Cambridge University Press.

Büchs, M., & Koch, M. (2019). Challenges for the degrowth transition: The debate about wellbeing. *Futures, 105*, 155–165.

Chaney, D. (1996). *Lifestyles*. Routledge.

Climate Analytics and New Climate Institute. (2022). *The CAT thermometer*. Retrieved from Climate Action Tracker website: https://climateactiontracker.org/global/cat-thermometer/

Dew, N., Aten, K., & Ferrer, G. (2017). How many admirals does it take to change a light bulb? Organizational innovation, energy efficiency, and the United States Navy's battle over LED lighting. *Energy Research & Social Science, 27*, 57–67.

Dunlap, R. E., & Brulle, R. J. (2020). Sources and amplifiers of climate change denial. In D. C. Holmes & L. M. Richardson (Eds.), *Research handbook on communicating climate change* (pp. 49–61). Elgar Online.

Ekwurzel, B., Boneham, J., & Dalton, M. W. (2017). The rise in global atmospheric CO_2, surface temperature, and sea level from emissions traced to major carbon producers. *Climatic Change, 144*, 579–590.

Ford, R., Walton, S., Stephenson, J., Rees, D., Scott, M., King, G., Williams, J., & Wooliscroft, B. (2017). Emerging energy transitions: PV uptake beyond subsidies. *Technological Forecasting and Social Change, 117*, 138–150.

Geels, F. W. (2002). Technological transitions as evolutionary reconfiguration processes: A multi-level perspective and a case-study. *Research Policy, 31*(8–9), 1257–1274.

Geels, F. W., Pieters, T., & Snelders, S. (2007). Cultural enthusiasm, resistance and the societal embedding of new technologies: Psychotropic drugs in the 20th century. *Technology Analysis and Strategic Management, 19*(2), 145–165.

Geels, F. W., & Verhees, B. (2011). Cultural legitimacy and framing struggles in innovation journeys: A cultural-performative perspective and a case study of Dutch nuclear energy (1945–1986). *Technological Forecasting & Social Change, 78*(6), 910–930.

Giddens, A. (1984). *The constitution of society: Outline of the theory of structuration*. University of California Press.

Goggins, G., Rau, H., Moran, P., Fahy, F., & Goggins, J. (2022). The role of culture in advancing sustainable energy policy and practice. *Energy Policy, 167*(May), 113055.

Griffin, P. (2017). *The carbon majors database: CDP carbon majors report 2017*. CDP & Climate Accountability Institute.

Hawkins, G., & Muecke, S. (Eds.). (2003). *Culture and waste: The creation and destruction of value*. Rowman & Littlefield.

Hays, S. (1994). Structure and agency and the sticky problem of culture. *American Sociological Association, 12*(1), 57–72.

Hopkins, D., & McCarthy, A. (2016). Change trends in urban freight delivery: A qualitative inquiry. *Geoforum, 74*, 158–170.

Hopkins, D., & Stephenson, J. (2014). Generation Y mobilities through the lens of energy cultures: A preliminary exploration of mobility cultures. *Journal of Transport Geography, 38*, 88–91.

Hopkins, D., & Stephenson, J. (2016). The replication and reduction of automobility: Findings from Aotearoa New Zealand. *Journal of Transport Geography, 56*, 92–101.

International Energy Agency. (2020). *Sustainable recovery*. World Energy Outlook Special Report.

IPBES. (2019). *Global assessment report on biodiversity and ecosystem services of the Intergovernmental Science-Policy Platform on Biodiversity and Ecosystem Services* (E. S. Brondizio, J. Settele, S. Díaz, & H. T. Ngo, Eds.). IPBES secretariat.

IPCC. (2022). *Climate change 2022: Mitigation of climate change. Contribution of Working Group III to the sixth assessment report of the intergovernmental panel on climate change*. Intergovernmental Panel on Climate Change.

Jackson, T. (2021). *Post growth: Life after capitalism*. Polity Press.

Johnson, O. W., Gerber, V., & Muhoza, C. (2019). Gender, culture and energy transitions in rural Africa. *Energy Research & Social Science, 49*, 169–179.

Jürisoo, M., Serenje, N., Mwila, F., Lambe, F., & Osborne, M. (2019). February). Old habits die hard: Using the energy cultures framework to understand drivers of household-level energy transitions in urban Zambia. *Energy Research and Social Science, 53*, 59–67.

Kallis, G. (2018). *Degrowth*. Agenda Publishing.

Kartha, S., Kemp-Benedict, E., Ghosh, E., Nazareth, A., & Gore, T. (2020). *The Carbon Inequality Era: An assessment of the global distribution of consumption emissions among individuals from 1990 to 2015 and beyond*. Stockholm Environment Institute & Oxfam.

Klein, N. (2015). *This Changes Everything*. Penguin.

Lade, S. J., Steffen, W., De Vries, W., Carpenter, S. R., Donges, J. F., Gerten, D., Hoff, H., Newbold, T., Richardson, K., & Rockström, J. (2020). Human impacts on planetary boundaries amplified by Earth system interactions. *Nature Sustainability, 3*(2), 119–128.

Latour, B. (1996). On actor-network theory: A few clarifications. *Soziale Welt, 47*(4), 369–381.

Lawson, R., & Williams, J. (2012). Understanding energy cultures. In *Proceedings of the Annual Conference of the Australia and New Zealand Academy of Marketing (ANZMAC)*. University of New South Wales.

Manouseli, D., Anderson, B., & Nagarajan, M. (2018). Domestic water demand during droughts in temperate climates: Synthesising evidence for an integrated framework. *Water Resources Management, 32*(2), 433–447.

McCauley, D., & Heffron, R. (2018). Just transition: Integrating climate, energy and environmental justice. *Energy Policy, 119*, 1–7.

McCracken, G. D. (1990). *Culture and consumption: New approaches to the symbolic character of consumer goods and activities*. Indiana University Press.

McKague, F., Lawson, R., Scott, M., & Wooliscroft, B. (2016). Understanding the energy consumption choices and coping mechanisms of fuel poor households in New Zealand. *New Zealand Sociology, 31*(1), 106–126.

Midgely, G. (2003). *Systems thinking*. Sage.

Ortner, S. B. (1984). Theory in anthropology since the sixties. *Comparative Studies in Society and History, 26*(1), 126–166.

Parrique, T., Barth, J., Briens, F., Kerschner, C., Kraus-Polk, A., & Kuokkanen A., Spangenberg, J. H. (2019). *Decoupling debunked: Evidence and arguments against green growth as a sole strategy for sustainability*. European Environmental Bureau.

Peace, A., Connor, L. H., & Trigger, D. (2012). Environmentalism, culture, ethnography. *Oceania, 82*(3), 217–227.

Piketty, T. (2020). *Capital and ideology*. Harvard University Press.

Quinn, E., & Westwood, B. (Eds.). (2018). *Thinking veganism in literature and culture: Towards a vegan theory*. Palgrave Macmillan.

Raworth, K. (2017). *Doughnut economics: Seven ways to think like a 21st-century economist*. Chelsea Green Publishing.
Reckwitz, A. (2002). Toward a theory of social practices: A development in culturalist theorizing. *European Journal of Social Theory, 5*(2), 243–263.
Scott, M. G., & Lawson, R. (2018). The road code: Encouraging more efficient driving practices in New Zealand. *Energy Efficiency, 11*(7), 1617–1626.
Scott, M. G., McCarthy, A., Ford, R., Stephenson, J., & Gorrie, S. (2016). Evaluating the impact of energy interventions: home audits vs. community events. *Energy Efficiency, 9*(6), 1221–1240.
Shove, E., & Spurling, N. (Eds.). (2013). *Sustainable practices: Social theory and climate change*. Routledge.
Sovacool, B. K., & Griffiths, S. (2020a). Culture and low-carbon energy transitions. *Nature Sustainability, 3*(9), 685–693.
Sovacool, B. K., & Griffiths, S. (2020b). The cultural barriers to a low-carbon future: A review of six mobility and energy transitions across 28 countries. *Renewable and Sustainable Energy Reviews, 119*, 109569.
Spangenberg, J. H. (2017). Hot air or comprehensive progress? A critical assessment of the SDGs. *Sustainable Development, 25*(4), 311–321.
Steffen, W., Richardson, K., Rockström, J., Cornell, S. E., Fetzer, I., Bennett, E. M., Biggs, R., Carpenter, S. R., De Vries, W., De Wit, C. A., Folke, C., Gerten, D., Heinke, J., Mace, G. M., Persson, L. M., Ramanathan, V., Reyers, B., & Sörlin, S. (2015). Planetary boundaries: Guiding human development on a changing planet. *Science, 347*(6223), 1259855.
Stephenson, J. (2018). Sustainability cultures and energy research: An actor-centred interpretation of cultural theory. *Energy Research & Social Science, 44*(May), 242–249.
Stephenson, J. (2020). Sustainability cultures: Exploring the relationships between cultural attributes and sustainability outcomes. In *The Cambridge handbook of environmental sociology* (pp. 236–248). Cambridge University Press.
Stephenson, J., Barton, B., Carrington, G., Doering, A., Ford, R., Hopkins, D., Lawson, R., McCarthy, A., Rees, D., Scott, M., Thorsnes, P., Walton, S., Williams, J., & Wooliscroft, B. (2015). The energy cultures framework: Exploring the role of norms, practices and material culture in shaping energy behaviour in New Zealand. *Energy Research & Social Science, 7*, 117–123.
Stephenson, J., Barton, B., Carrington, G., Gnoth, D., Lawson, R., & Thorsnes, P. (2010). Energy cultures: A framework for understanding energy behaviours. *Energy Policy, 38*(10), 6120–6129.
Stephenson, J., Barton, B., Carrington, G., Hopkins, D., Lavelle, M. J., Lawson, R., Rees, D., Scott, M., Thorsnes, P., Walton, S., & Wooliscroft, B. (2016). *Energy Cultures Policy Briefs*. Centre for Sustainability, University of Otago.

Stephenson, J., Hopkins, D., & Doering, A. (2015). Conceptualizing transport transitions: Energy Cultures as an organizing framework. *Wiley Interdisciplinary Reviews: Energy and Environment, 4*(4), 354–364.

Stephenson, J. R., Sovacool, B. K., & Inderberg, T. H. J. (2021). Energy cultures and national decarbonisation pathways. *Renewable and Sustainable Energy Reviews, 137*, 110592.

United Nations. (2015). *Transforming our world: The 2030 agenda for sustainable development*.

United Nations. (2022). *The sustainable development goals report 2022*.

Urry, J. (2004). The 'system' of automobility. *Theory, Culture & Society, 21*(5), 25–39.

Walton, S., Doering, A., Gabriel, C.-A., & Ford, R. (2014). *Energy transitions: Lighting in Vanuatu*. Project report, Australian Aid.

Walton, S., Zhang, A., & O'Kane, C. (2020). Energy eco-innovations for sustainable development: Exploring organizational strategic capabilities through an energy cultures framework. *Business Strategy and the Environment, 29*(3), 812–826.

Williams, R. (1976). *Keywords: A vocabulary of culture and society*. Oxford University Press.

Wilson, S., Carlson, A., & Szeman, I. (Eds.). (2017). *Petrocultures: Oil, politics, culture*. McGill-Queen's University Press.

Winkler, H. (2020). Towards a theory of just transition: A neo-Gramscian understanding of how to shift development pathways to zero poverty and zero carbon. *Energy Research & Social Science, 70*, 101789.

Young, W., & Middlemiss, L. (2012). A rethink of how policy and social science approach changing individuals' actions on greenhouse gas emissions. *Energy Policy, 41*, 742–747.

CHAPTER 2

Culture's Divergence

Introduction

Culture is a difficult yet promising concept to apply to sustainability challenges because of its diverse interpretations. It has a variety of meanings in everyday language, various interpretations depending on disciplines and their theoretical orientations, and different fundamental conceptualisations in Western and Indigenous worldviews. This chapter focuses on explaining these divergences because unless they are visible and named, they will continue to handicap the effective use of culture as a lens for examining sustainability issues. The latent promise of this divergence, which will be clear by the end of this book, lies in the rich bodies of knowledge that underpin culture's many interpretations and their potential contribution to sustainability transitions.

Culture's meaning has been evolving since it first appeared in the English language over 500 years ago, and there have been many attempts to achieve some sense of order across different understandings of culture: either via its etymological evolution (e.g. Bennett et al., 2005; Williams, 1976) or through reviewing academic definitions of culture (e.g. Faulkner et al., 2006; Kroeber & Kluckhohn, 1952) and/or by grouping similar academic applications (e.g. Hammersley, 2019). Standard dictionaries of English typically offer three groups of common usages: one referring to the arts and other manifestations of intellectual and skilful achievement; a second referring to ways of life of a particular group of people; and a third referring to the cultivation of plants and animals, including cells and

J. Stephenson, *Culture and Sustainability*,
https://doi.org/10.1007/978-3-031-25515-1_2

micro-organisms. Williams (1976) identified three main meanings: one aligning with artistic activity and works of art; the second referring to the process of societal development to an intellectual, spiritual or aesthetic ideal; and the third broadly referring to different ways of life. Forty-five years later, a review confirmed these three usages and added a fourth: culture as the process and outcomes of shared meanings (Hammersley, 2019).

My own review, which occupies most of this chapter, identifies nine senses in which culture is used. I focus on different perspectives of culture in the light of their potential to be applied to sustainability problems. Eight clusters of meaning derive from its use in Western language and scholarship. They include Williams' and Hammersley's categories and also draw from culture's past and more recent usages. I also describe a further meaning that is implicit in many non-Western worldviews. These nine clusters are hard evidence, if any more were needed, of why culture can be such a slippery term.

Literature on culture is extensive and theoretically complex. In this chapter, I have attempted to explain key differences in interpretations in the simplest language possible so that it makes sense to a broad audience. I am not able to do justice to these rich and sometimes hotly argued fields of knowledge. For those who are interested in diving more deeply into any of these areas of literature, I have included some key citations as a starting point.

Culture's Multiple Meanings

Culture's divarication started early. Its linguistic origins in English are from the early fifteenth century where it initially referred to the tending of crops. It evolved to span several other meanings (human development; the arts; folk cultures) before it began to be used academically in the nineteenth century (Williams, 1976). Its meanings diverged further as it became a key concept in many social science and humanities disciplines, developing more complex and more nuanced meanings than are evident in standard dictionary definitions. From being a core focus of anthropology (initially applied to non-Western cultures) and humanities (relating to works of aesthetic and intellectual endeavour), it became more widely adopted across social science and humanities disciplines from the early 1970s. This 'cultural turn' was largely an outcome of the rise in academic interest in meaning and symbolism (of actions, objects, texts, discourses, etc.), which was a reaction against the previously dominant focus on empirically observable 'facts' (Chaney, 1994). This shift in focus resulted in new subdisciplines, such as cultural geography, cultural archaeology, cultural sociology, sociology of culture, cultural psychology, cultural history and the general field of cultural studies, each with its own particular take on culture. Ongoing waves of 'posts', 'turns' and 'isms' (e.g. post-structuralism, the performative turn, new materialism) have continued to influence culture's interpretations and have opened up new avenues for cultural scholarship.

Alongside these ever-evolving interpretations has been an ongoing expansion in culture's applications as a descriptor. Where in past decades people spoke of native cultures, folk cultures, high culture and mass culture, they are now more likely to talk about consumer culture, drug cultures, gamer cultures, visual culture, organisational cultures and cancel culture. Culture is applied, with one or several of its meanings, to define or explain a bewildering and shifting array of social experiences.

As a result of these divergences, culture has accrued a range of meanings and finely wrought nuances such that different fields of knowledge can sit in 'baneful isolation' (Patterson, 2014: 3). Academics writing about culture generally do so based on their specialised interpretation, supported by the annexation of everyday words (such as 'practice', 'performance', 'text' or 'structure') to mean something very specific to that discipline. The result can be dense and impenetrable prose, creating barriers for communication even for academics in other social science or humanities disciplines, and making potentially enlightening cross-fertilisations unlikely. For academics in other disciplines and for lay people, academic writing on culture can be wholly inaccessible.

The combination of divergent meanings, ever-expanding applications and conceptual–linguistic specialisation means that culture can be highly ambiguous, such that 'what counts as culture depends upon what is being described or explained, and for what purposes' (Hammersley, 2019: 96). The ambiguity becomes particularly problematic when academics and students who are not based in a cultural discipline use the term 'culture' and fail to define what they mean, appearing to assume that their readers interpret the term as they do, just as we would with any commonly used noun. This 'persistent lack of consensus or rigor in defining culture' (Patterson, 2014: 3) means that it can be subject to misinterpretation and hence any claims may be of questionable value to others.

A further problem with culture lies in its Eurocentric origins and meanings. Non-Western societies, and particularly Indigenous societies, have very different worldviews and knowledge systems, and these are reflected in language. Even where an Indigenous word may superficially appear to be equivalent to a term in English, it will be foundationally different and thus convey a set of meanings to native language speakers that are not evident to others. Indigenous and other societies may conceive of culture in fundamentally different ways or may use a seemingly equivalent word that has subtle or substantial differences in meaning. In my view, this is not a reason to dismiss non-Western conceptualisations, but rather an opportunity to open up new possibilities for understanding the world.

Clearly, then, the scope of 'what culture means' is very broad. As a concept, it cannot do useful work for the sustainability transition unless we can articulate its scope and are able to consider how different interpretations relate to each other. In the following sections, I discuss the nine main clusters of meaning that I have derived from a broad review of literature relating to culture. Across these clusters, it will be seen how the concept of culture has diverged both ontologically (the nature of the reality that culture represents) and epistemologically (the kinds of evidence or knowledge that can describe that reality).

Some of the clusters of meaning (numbers 1, 3 and 4 below) will be unsurprising to the lay reader as they closely align with everyday understandings, so require little explanation. One cluster (number 2 below) is largely obsolete in its original application but has a different relevance today. Other clusters (numbers 5–8 below) emerge out of developments in the social sciences, and in these instances, I spend a little longer outlining relevant theories so that the meaning applied to culture makes sense to a lay reader. For non-Western understandings of culture (number 9 below), I focus in particular on Indigenous perspectives and draw from literature relating to a small number of Indigenous societies. I am unable to do this topic full justice, because in reality there are potentially thousands of different non-Western perspectives, and no reason to think that they do or must align. My intention is simply to open the door to the possibilities for entirely different interpretations, worldviews and language systems to contribute to understandings of culture.

Culture-As-Nurture

Culture's original meaning was akin to husbandry, referring to the tending, caring and cultivation of plants and animals (Williams, 1976). This root meaning is carried on today in words like 'agriculture' and 'horticulture' and has a specialised application to laboratory-based 'cultures' where it refers to artificially maintained cells or bacteria (Bennett et al., 2005). Although absent from most social science discussions of culture, I believe this root meaning has important implications for sustainability.

Recent decades have seen the widespread growth of industrial agriculture. This involves increasingly large farming operations, crop and animal monocultures, heavy use of fertilisers, herbicides and pesticides, large irrigation schemes and mechanised farm processes. These operations often replace smaller and more diverse farms or are established on previously forested land. Impacts include increased greenhouse gas emissions, habitat and species losses, pollution of water bodies and water scarcity, with consequential societal impacts including increasing inequalities, debilitating health impacts and food insecurity (IPES-Food, 2018; United Nations Environment Programme, 2021).

Global bodies such as the United Nations and the Food and Agriculture Organisation are increasingly calling for the transformation of agriculture away from the industrial model to return to food and farming systems where the focus is not just on producing the desired outputs but on nurturing the natural systems that support their production. Traditionally, communities in most parts of the world tended local food systems in ways that resulted in unique eco-cultural systems that simultaneously supported flourishing local ecosystems and human communities, and these have the potential to be restored (Koohafkan & Altieri, 2016). Other more modern forms of sustainable farming include agroecology or regenerative agriculture, which seek to de-intensify farming and develop mutually nurturing interactions between plants, animals, humans and the environment (Burns, 2020).

In all of these latter instances, the 'culture' in agriculture is returning closer to its original meaning of tending, caring and nurturing. This requires a cultural shift among farmers, in a wider sense of the word. Regenerative farming, for example, has been described as a social movement as much a change to farm techniques, involving alterations in 'beliefs, values, emotions, worldviews, structures of meaning-making, and consciousness more generally' (Gosnell et al., 2019: 1). To turn around the destructive effects of industrialised agriculture on climate, water, ecologies, health and livelihoods will involve reclaiming the 'culture' of agriculture and horticulture to more of its original sense: tending not only crops and animals but also the ecological and social systems with which they are mutually enmeshed.

Interpreting culture as nurture invites questions such as: *What does a more nurturing form of agriculture look like?* and *What can we learn from farmers who already tend their farms sustainably?*

Culture-As-Progress

From the seventeenth century, culture became a metaphor for human development, as in culturing the mind towards an ideal state of existence. By the late nineteenth century, through various linguistic and social influences, this had become associated with the idea that there was a pinnacle of civilisation towards which human societies should be evolving (Williams, 1976). Unsurprisingly, this civilisation was described in terms of European intellectual, spiritual and aesthetic priorities.

This conceptualisation of culture has had devastating effects because it underpinned colonial and imperialist activities, with Western nations partly justifying their actions on the basis that they were bringing 'civilisation' to the rest of the world (Ferguson, 2012). Similarly, within the emerging discipline of anthropology, early studies of 'primitive cultures' took the view that there was a natural progression through which societies evolved to become more advanced or civilised, aligning with the perceived superiority of Western culture (Hammersley, 2019). These notions are now widely condemned due to their role in hegemonies of power, racism and inequality.

At the same time, culture's meaning took on an anti-progress dimension as a result of the industrial revolution. Industrialisation saw a mass movement of rural people to cities, despoilation of natural areas through mining and industrial development, and new forms of work and social organisation. Many of the intellectual elite were appalled at the impacts on traditional ways of life. The newly urbanised societies were unfavourably compared with English and European rural 'folk' cultures (Hammersley, 2019). From this perspective, culture referred to traditional ways of life that were being lost as a result of modernity.

Culture is still used in the sense of an ideal set of social qualities. The harmful consequences of this interpretation are visible from the personal scale of being an outcast from a desirable social group, to societies in which displaying difference is dangerous, to wars of aggression that seek to destroy or subjugate those with different beliefs, practices or languages. From a sustainability perspective, it also is visible in socio-technical imaginaries that idealise modernity and its reliance on fossil fuels, consumption and growth (Jasanoff & Kim, 2015; Stoddard et al., 2021).

Despite its negative connotations, culture-as-progress can be spun another way, so that rather that describing an idealised set of cultural attributes, it describes an ideal set of outcomes. If 'progress' is redefined as cultural change towards more sustainable outcomes, it means valuing multiple different approaches to sustainability. Used in this way, there is no blueprint for a 'perfect' set of cultural arrangements. Cultural diversity becomes all-important, a topic that I will return to later in the book.

This perspective invites questions such as: *How do we know when a culture is sustainable? What specific cultural features (norms, beliefs, practices,* etc.*) align with more sustainable outcomes? What can we learn from cultures that have established ways of living sustainably? What process of change do we see in cultures that become more sustainable over time?*

Culture-As-Product

From the late eighteenth century, culture began to be used to refer to products of the idealised conception of Western civilisation described above. Poet and literary critic Matthew Arnold was particularly influential in adopting culture to refer to the best of what has been known and said in the world (Collini, 1993). From this perspective, culture referenced particular types of aesthetic and intellectual products—literature, art, music, drama and ideas—that epitomised the pinnacle of civilisation. This was the start of culture becoming closely associated with the arts and with upper and middle classes.

Culture's use in this sense remained elitist until the twentieth century, when the idea of there being a monopoly on culture as a normative standard started to erode. New forms of media enabled the proliferation of movies, music and other art forms (some of which directly challenged elitism) and this led to concepts of 'mass culture' or 'popular culture'. Cultural theorists introduced the term 'culture industry' to describe that part of the capitalist system that produces cultural artefacts for mass consumption and to support its own replication (Adorno & Horkheimer, 1944). Alongside this, artistic outputs from working classes and non-Western societies began to be more widely recognised and appreciated.

Today, culture in this sense is used to refer to an extraordinarily broad range of products and practices including artworks, music, theatre, dance, visual media, fashion, online content, and structures and places that represent shared values, meanings or preferences. The vestiges of culture as 'high art' still remain (e.g. referencing classical music, artworks and drama) but are largely eclipsed by its democratisation: today almost anyone can produce cultural artefacts, and cultural products can be almost anything, as long as others recognise it as such.

This interpretation of culture has made its way into the concept of 'cultural capital'. As initially introduced by Pierre Bourdieu, it referred to people's familiarity with 'high culture' and their ability to use this knowledge to their advantage in their social lives (Bourdieu, 1984). Cultural capital is now used more broadly to refer to familiarity with cultural products, uses of language and symbolism, ideas, tastes and preferences, along with skills and knowledge that can be strategically put to use to further one's interests (Hanquinet & Savage, 2015).

From a sustainability perspective, culture-as-product is critically important because cultural outputs are highly influential in shaping how people perceive the world. For example, dominant forms of cultural media (radio, television, movies and latterly the digital world) have long encouraged unsustainable activities by normalising lifestyles and business operations that value over-consumption, greed and individualism. These ideas have permeated many societies to the extent that they have become almost unquestionable, embedded as invisible statements of 'normal'. In contrast, societies that are not so embedded in Western cultural ideals can convey deeply different perspectives through their art, stories and other cultural outputs (Yunkaporta, 2020).

While there is already considerable activity in the arts supporting a sustainable future, this is still a relatively minor segment of cultural production. Culture-as-product invites a focus on questions such as: *How can we create new visions of a sustainable future? How can art help reveal and challenge unsustainable practices? How can we build cultural capital that empowers people to take an active part in the sustainability transition?*

Culture-As-Lifeways

In the mid-late nineteenth century, culture began to be used as a descriptor of the entire way of life of any group of people. Initially, this was predominantly an academic concept and emerged as a reaction against the dominance of Western cultural ideals as reflected in culture-as-progress and culture-as-product. An early influence was the work of German anthropologist Franz Boas, who argued that the evaluative hierarchy implied by culture-as-progress was socially harmful and that all societies set their own expectations of value (Bennett et al., 2005).

Rather than being aligned with a narrow interpretation of civilisation, culture started to be applied to the distinctive ways in which groups of people lived. Anthropological studies explored tangible aspects of culture such as clothing, foods and tools, along with practices and rituals, and sought to understand the meanings, knowledge and beliefs that were shared within a given social group. An early anthropological definition of culture in this sense was 'that complex whole which includes knowledge, belief, art, law, morals, custom, and any other capabilities and habits acquired by man [sic] as a member of society' (Tylor, 1920: 1).

The exploration of culture-as-lifeways became central to the discipline of anthropology from the 1920s. Investigations were initially of non-Western communities or ethnicities, often with a focus on societal structures like marriage, kinship and methods of exchange as well as the patterns of everyday life (Ortner, 1984). From the mid-twentieth century, anthropologists (and, later, sociologists) began to take an interest in cultures in the Western world, including diverse but distinctive groups such as youth cultures, gay cultures, sports cultures and other so-called subcultures and counter-cultures (Hammersley, 2019). Cultural analyses included religious, occupational or recreational aspects of everyday life.

Later in the twentieth century, culture became considered more generally as a system of interrelated characteristics centred on shared meanings, as reflected in this definition: 'not only the beliefs and values of social groups, but also their language, forms of knowledge, and common sense, as well as the products, interactional practices, rituals, and ways of life established by these' (Hays, 1994: 65). Language in this sense includes the subtle selections of words, phrases and patterns of speech that are shared by a group, including speech styles, terms, dialects and jargon. Common sense refers to shared understandings about what is 'normal' and largely unquestionable to that group. Culture-as-lifeways acknowledges the tangible features of culture but also encompasses its more intangible qualities such as shared meanings, beliefs and values.

Today, culture-as-lifeways continues to be explored academically through many disciplinary frames including in sociology, geography and anthropology. It is now broadly accepted that many people in the modern world 'live culturally' rather than within a particular culture. People shift between cultures in different parts of their lives, moving relatively seamlessly between varied cultural expectations (e.g. between home, work and sport, and depending on who they are with). At the same time, cultural phenomena that are passed down through generations and across population (e.g. ideas, norms, practices, physical objects) are influential in maintaining continuities over time and space, and are thus important (but not sole) causal influences in societal trajectories (Patterson et al., 2004).

The ways in which people live—their everyday activities, actions, acquisitions, tools, products, beliefs, norms and values—clearly have implications for sustainability. Some ensembles of cultural features will have better sustainability outcomes than others. If we were using carbon emissions as a measure of sustainability, for example, people who eat little meat, rarely fly and always use active and shared transport will have fewer carbon emissions that those who have a meat-based diet and are heavy users of private cars and air travel. If ecological footprint was the measure of sustainability, the cultures that value conspicuous consumption will likely have far higher ecological impacts than those that have preferences for more frugal living. Cultures that favour individual rights over collective welfare tend to have much higher levels of income disparity. Different ways of living have direct implications for ecological, social and economic outcomes.

A way-of-life perspective on culture invites explorations of the causal links between cultural attributes and sustainability. It invites questions such as: *What ways of living offer the best chance of achieving a sustainable future? What adjustments to beliefs, expectations, rules of behaviour, habitual practices and material possessions might be required? Can we gain insights into more sustainable lifestyles from cultures other than our own?*

Culture-As-Meaning

From the 1970s, the academic study of culture underwent a distinct swing and began to focus intently on shared meanings and symbolism. Influences on this intellectual shift included the work of Clifford Geertz, Pierre Bourdieu and Michel Foucault, who explored how shared meanings underpin social processes such as social cohesion, power and exclusion (Bourdieu, 1977; Foucault, 1978, 1980; Geertz, 1973). From this perspective, culture was a set of ideologies and other control mechanisms for the governing of behaviour, incorporating 'systems of meaning in terms of which we give form, order, point, and direction to our lives' (Geertz, 1973: 13).

Culture in this sense comprises the invisible world of shared ideas, symbolism, feelings, beliefs and values. It usually does not include the empirically observable traces of these cognitive processes: 'what people actually do, how they behave, the institutions they construct, and the physical exchanges of money and power in which they engage, however, are not part of culture' (Wuthnow et al., 1984: 4). Objects or texts or actions are studied for the shared meanings that they convey, how they affect people cognitively and emotionally, and their implicit messages (e.g. of power, class, exclusivity and belonging) (Kasanga, 2015).

Culture-as-meaning was widely adopted across the social sciences and humanities, leading to new sub-disciplinary branches within anthropology, sociology and geography, and its widespread adoption in media studies. A new interdisciplinary grouping known as cultural studies, influenced by Marxism, took a particular interest in power and ideology (During, 1999). These branches diverged further on points of theory, such as whether the social world consists only of shared meanings or includes tangible phenomena, and whether the process of meaning-making is conscious or occurs below the level of conscious awareness (Hammersley, 2019). Cultural theories of meaning can be grouped into three main fields: those that identify the source of shared meanings in symbol-conveying phenomena such as objects, texts or discourses; those that identify the source of shared meaning in symbols held in people's minds; and those that identify it in the ways in which people interact and communicate with one another (otherwise known as culturalist textualism, culturalist mentalism and intersubjectivism [Reckwitz, 2002]). Culture-as-meaning also gave rise to new methodological approaches that sought to interpret discourses, texts, images and mental states to identify their underlying meanings and interpolate how these shape social phenomena such as ideologies, institutions and social actions.

Culture-as-meaning has potentially strong relevance to sustainability questions. Unsustainable systems of production and consumption are underpinned by ideologies, shared meanings and symbolism, often messaged overtly through advertising as well as covertly through alignment with social influencers and other mechanisms; messaging that locks consumers into wanting more (Heath & Chatzidakis, 2012). Under standard capitalist models, corporations are expected to be self-serving agents focused on ongoing growth and wealth, and these ideologies permit worker exploitation and environmentally destructive resource extraction and manufacturing. To achieve transitions away from such models, it is critical to reveal their underlying ideologies and symbolisms, and to start to replace them with new meanings that align with sustainable products, services and ways of life. The start of such a shift is visible in the increasing value given by consumers to sustainable products and services, and in businesses that fundamentally realign their values and purpose with sustainability objectives, but we still have far to go.

Culture-as-meaning invites us to consider questions such as: *How can we make visible the problematic ideologies/belief systems on which overconsumption is based? What new meanings and symbols convey sustainability concepts, and what is their impact in the social world? To what extent is the concept of sustainability developing its own ideologies and meanings, and are these consistent with a truly sustainable future?*

Culture-As-Structure

Structures (also called 'social structures') are generally considered to be enduring patterns of social relationships that are underpinned by shared values and ideologies (Hays, 2000; Patterson, 2014). They include 'capitalism, bureaucracy, the state, social networks, social classes, status groups, population dynamics, and the distribution of material resources' (Hays, 2000: 597). Some aspects of structure, such as shared rules, values and ideologies, sound not dissimilar to culture, and I will return to this point.

Social structure has been central to the study of society since the mid-1800s, with early contributions by Karl Marx, Herbert Spencer, Max Weber, Ferdinand Tonnies and Emile Durkheim. Studies of structure are broad ranging, including analyses of the structures of capitalism, structure through a feminist lens and the structural foundations of culture (Browne, 2005). In different ways, all identify powerful formal and informal shared 'rules', seemingly independent of individuals, which constrain human choices and shape their actions. Social stability exists because people agree to (or are forced to abide by) these shared expectations of behaviour.

A highly influential theoretical development in sociology in the twentieth century was Anthony Giddens' theory of structuration. This offered an explanation of why social structures endure, and why, despite this, human choices are not predetermined (Giddens, 1984). He proposed that structure is simply the routinised actions of many people over time, according to collective 'rules' or expectations about behaviour. Structures are rules that people tend to follow, and the rules exist because people continue to enact them. He called this two-way process the duality of structure: 'the structural properties of social systems are both the medium and the outcome of the practices that constitute those systems' (Giddens, 1979: 69). Sewell elaborates on Giddens' 'rules', describing them as cultural schemas which are 'taken-for-granted mental assumptions or modes of procedure that actors normally apply without being aware that they are applying them' (Sewell, 1992: 22).

At the same time, people have a degree of freedom in how they act, which Giddens terms 'agency'. Choices outside of the rules can occur but are generally infrequent, and thus people's actions generally replicate social structures (Scott, 2007). The ability to make choices that defy structures is largely dependent on the degree to which individuals or groups have what Giddens calls 'resources'. These include personal or collective capabilities and capacities, as well as physical manifestations such as property and wealth, both of which equate to power. Agency is constrained where people have little power and enhanced where people have more power. I will return to agency in Chapter 4.

In many senses, structure and culture describe the same (or similar) sets of attributes: they are both concerned with norms, beliefs, values, principles and conventions that shape practices and behavioural tendencies, as well as the resources that actors work with, acquire, produce and accumulate. Some authors differentiate between culture and structure by using culture to refer to a 'soft' system of subjective values, meanings and symbolism (e.g. Hannerz, 1992; Parsons, 1951), while social structures are 'hard' institutionalised social arrangements that are more permanent and pervasive such as economic systems, kin relations and political institutions (Bernandi et al., 2006; Hays, 1994). Capitalism, as an example, is described as 'a spectacular case of a power-laden yet long-enduring structure' (Sewell, 1992: 25).

Some writers suggest the difference is scale and persistence, with structure being more durable and less ephemeral than culture (Lentz, 2017). But others argue that structure can be analysed at many scales. These include long-term globally pervasive structures such as capitalism, to mid-scale and mid-term structures such as might exist with a political regime, through to localised and shorter-term structures that exist in specific personal or organisational relationships; sometimes referred to as macro, meso and micro level structures (Bernandi et al., 2006). Culture can similarly be seen to operate at different scales, from the ways of life and meanings shared by small groups of people to widely shared ideologies.

Unsurprisingly, some scholars suggest that culture and structure are indistinguishable. For many theorists (especially anthropologists), culture *is* the structure that orders social life. Giddens himself noted that his concept of structure is what many would refer to as culture (Scott, 2007). Both structure and culture refer to shared meanings and rules, and the manifestations of these in discernible patterns of practices and material possessions. While social structure is often treated as analytically distinct from culture, the two terms are arguably referring to the same sets of characteristics and the same processes. Sociologist Sharon Hays argues that 'the theoretical misstep of separating "structure" and "culture" is one of the principal conceptual problems that keeps sociologists from recognizing the power and centrality of culture' (Hays, 2000: 597).

Although structure and culture may be largely indistinguishable, the concept of structure usefully highlights the existence of pervasive and enduring cultural phenomena: 'the deepest, most recurrent aspects of social reality, its framework or underlying form' (Bernandi et al., 2006: 162). By including structure as one of the interpretations of culture, I am drawing attention to the literature that explores these deeper manifestations, and the insights that culture-as-structure can offer to sustainability questions.

From a sustainability perspective, the transition will not be achieved without fundamental changes to culture-as-structure. Deeply held ideologies—such as the growth paradigm, the right to benefit from the destruction of natural systems, and the sole responsibility of business to make profit for its shareholders—underpin the dominant economic system and its unsustainable outcomes. Recent societal shifts that are responding to sustainability crises, such as the growing acceptance of the concept of degrowth, discussions of the rights of nature and new forms of corporate responsibility, may be the beginning of fundamental changes in culture-as-structure.

Both cultural and structural traditions of scholarship can assist in transforming culture-as-structure. Seeing structures as particular types of cultural manifestations opens up structure to investigation with tools of cultural analysis. It helps reveal how ideologies are interwoven with certain social systems, institutions, practices and materialities. Seeing long-established and influential forms of culture as structural turns the spotlight onto beliefs and power relations that are so deeply embedded and widely shared that they are almost invisible to adherents.

Culture-as-structure invites questions such as: *What are the fundamental structures that underpin and enable unsustainable systems? What ideologies, institutions and power relations are most responsible for systems that exploit people and planet? What macro and meso level structural changes will be required for a sustainable future, and where are those changes already under way?*

Culture-As-Practice

Social practice theory is a form of cultural theory that focuses on practices as the main way in which the social world is constituted. Its historical roots lie Giddens' structuration and its antecedents, as discussed under culture-as-structure, and also in the work of Pierre Bourdieu (Bourdieu, 1977). Giddens proposed that people undertake routine behaviours that align with shared formal or informal rules, which reinforce and reproduce the social realm. To help explain how social order is internalised in how people think and act, Bourdieu proposed the concept of 'habitus'—persistent patterns of thought, perceptions and action—which are themselves a response to the objective conditions within which the individual exists (Bourdieu, 1990). Habitus is acquired through the social and physical context of an individual and constrains the range of behaviours that an individual will feel comfortable to undertake.

Building on these concepts, practice theory takes the view that practices both create and evolve from shared meanings. Practices in this sense are habitually repeated activities that embody shared practical understandings. They are influenced by external rules and by people's desires, beliefs and expectations (Schatzki, 1997; Schatzki et al., 2001). Practice theory is related to the culture-as-meaning tradition, but differs significantly in its focus on bodily routines and the role of material objects as part of practices. In this latter aspect, practice theory draws inspiration from science and technology studies and its interest in the active role of 'things' in shaping human action and social processes (Almila, 2016; Shove et al., 2012).

Practice is more than simply a routine activity such as cooking or taking a bus; it is a 'routinized type of behaviour which consists of several elements, interconnected to one other: forms of bodily activities, forms of mental activities, "things" and their use, a background knowledge in the form of understanding, know-how, states of emotion and motivational knowledge' (Reckwitz, 2002: 249). Together, these make up a system of interconnected elements (a practice) which is considered to have its own agency. In this sense, practices historically precede individuals and 'recruit' them to take part, and individuals are hosts or carriers of practice (Røpke, 2009; Shove & Pantzar, 2007). Through their frequent repetition, practices reinforce and replicate themselves and the meanings they embody. Studies of practice have been as diverse as stock market trading, academic presentations, ironing, chess and Nordic walking.

Of all cultural theories discussed in this chapter, practice theory appears to be most widely used in sustainability-related research to date. Sustainability issues started to become dominant in practice theory research from around 2013, with areas of focus including consumer identity, sustainable consumption and household energy practices (Corsini et al., 2019). From a sustainability perspective, practice theory has mainly been used to investigate routine activities and their tendency for replication. It provides important insights into how social practices emerge and evolve, and why policies that rely on rational choice or information deficit models are unlikely to be successful (Shove, 2010, 2014). It has predominantly been used to explain social stability rather than social change, although it has also underpinned investigations into how everyday practices change over time (Shove et al., 2012), interventions for sustainability (Strengers & Maller, 2014) and the role of collective activity in large-scale socio-technical change (Welch & Yates, 2018).

Practice theory invites investigation of the role of habitual routines in cementing unsustainability. It invites questions such as: *Why are these unsustainable routines so widely shared and so hard to change? How can practice theory help explain the intractability of unsustainable business practices? How do practices change, and what stimulates a change in practices especially towards more sustainable outcomes?'.*

Culture-As-Purpose

In the late twentieth century, scholars started using a cultural lens to improve understandings of how businesses and other organisations operated. In management studies and the sociology of organisations, the concept of 'organisational cultures' became widely used, and methods were developed to analyse how organisational problems could be said to arise from its 'culture' (Wilson, 2001). Culture in this sense is variously defined, but generally covers shared values, belief and norms along with behaviours (Flamholtz & Randle, 2014).

A significant portion of papers in this field discuss the purposeful creation of sustainability-oriented organisational cultures. This includes cultures that support corporate sustainability (Galpin et al., 2015; Linnenluecke & Griffiths, 2010) and cultural change for more sustainable products and services (Matinaro & Liu, 2017; Obal et al., 2020; del Reyes-Santiago et al., 2017). This interest in the processes of intentional culture change for sustainability has expanded to include non-business organisations such as universities (e.g. Adams et al., 2018).

This work is in a similar vein to institutional theory, which seeks to understand why organisations behave similarly within their particular sectors. Here, rather than looking at culture within an organisation, the interest is in what might be called cultural convergence between businesses or organisations. Reasons for this 'institutional isomorphism' include normative pressures (i.e. from the standards and practices within the sector), mimetic pressures (i.e. imitation of successful businesses) or coercive pressures from the context within which the business operates (Daddi et al., 2020). These ideas align with how culture is learned by individuals—through copying others, through unwritten 'rules' of behaviour and from wider contextual influences. Some studies in this field have applied institutional theory to sustainability questions such as the uniform nature of corporate social responsibility and climate change reporting, and the growing adoption of climate change strategies (Comyns, 2018; Daddi et al., 2020).

What is particularly pertinent about all of this literature is the growing interest in purposeful change in cultures to achieve particular outcomes (Alvesson & Sveningsson, 2015). This is significantly different to most of the other approaches to culture discussed in this chapter where the primary focus is on cultural stability. In culture-as-purpose studies, culture is seen as something that can be intentionally created and changed.

Culture-as-purpose has the potential to be further developed as organisations, institutions, groups and communities intentionally seek to change aspects of their culture to achieve more sustainable outcomes. It invites investigation of questions such as: *What are key stimuli for organisational change towards sustainability? What are the dynamics of organisational cultural change? What conditions enable culture change within and among organisations? How can change in one organisation spread to influence the sector as a whole?*

Culture-as-Nature

In this final entry I reach beyond most Western thinking to suggest a further way that culture can be interpreted. As already touched on in Chapter 1, one of the most fundamental and widely influential concepts in Western thought is that culture is in distinct contrast to or is indeed the binary opposite of nature (Williams, 1980). To anthropologist Tim Ingold this 'single, master dichotomy [...] underpins the entire edifice of Western thought and science—namely that between the "two worlds" of humanity and nature' (Ingold, 2003: 22). This deeply engrained conceptual split situates nature (the physical environment and non-human life forms) 'out there', and culture (humans and their ideas, beliefs, knowledge, actions and the products of these) 'in here'.

This dualism has become so embedded in European language and thought that it is almost impossible to find a single word that expresses otherwise. Even among those working at the forefront of integrative thinking, the fundamental code of duality is still evident. Terms used to describe interconnections between people and their environments include 'ecological culture' (Escobar, 1996); 'co-evolving human and natural systems' (Gunderson & Holling, 2002); 'linked ecological and social systems' (Berkes et al., 2003); 'bio-cultural diversity' (Maffi, 2005); 'social-ecological systems' (Folke, 2006); 'sacred ecology' (Berkes, 1999); 'ethno-ecology' (Bridges & McClatchey, 2009); 'ethno-botany' (Grabherr, 2009); 'social-natural environments' (Duit et al., 2010); 'coupled human–environment systems' (Adger et al., 2010); and 'eco-cultures' (Pretty, 2011; Rapport & Maffi, 2011). These composite terms reflect the range of ways in which Western scholars are now starting to identify inseparable associations between people and their environments: ways in which they are 'mutually implicated in each others' coming into being' (Ingold, 2003: 306). Yet even though the purpose of each of these terms is to highlight the interactions and convergences between nature and culture, what is particularly notable is how the terms continue to replicate duality through the use of coupled words. This continued linguistic reinforcement of two realms, despite the clear intent of these scholars to do otherwise, reflects the difficulty of intellectually transcending the dominant dualism. If we are constrained by English language and its embedded thought patterns, the idea that culture can include nature might seem inconceivable—indeed, wrong.

Many Indigenous worldviews and language systems, on the other hand, reflect no such duality. When referring to Indigenous perspectives I will generally give examples from Māori, the Indigenous culture with which I am most familiar. Here I gratefully acknowledge the many conversations I have had with Māori friends, colleagues and community members over the past few decades which have helped me understand a little of tikanga Maori (customary system of values and practices). I also draw from academic literature by Indigenous scholars. I acknowledge the differences in tikanga that exist between Māori tribal groups, and also that Māori perspectives do not represent other Indigenous perspectives. However, there are some concepts that are widely shared by Māori and many other Indigenous societies, and one of these is that humans and nature are inseparable, and, indeed, that they are kin.

To Māori, for example, all things (including people) are descendants of the union of Ranginui, the sky father, and Papatūānuku, the earth mother. People are, therefore, genealogically linked to the earth and all living things (Mead, 2016). This inseparability is reflected linguistically. The word 'whenua', for example, means both 'land' and 'placenta', so one cannot conceive of land without considering at the same time one's relationship with it. The term 'whakapapa' refers to human ancestors but at the same time includes kin relationships with all aspects of the environment, including mountains, plants and animals (Roberts et al., 1995). A similar nature/culture unity is embedded in other Indigenous languages, such as the Fijian term 'vanua' (an area of land and sea considered as an integrated whole with its human occupants) and 'aschii/aski' (a living landscape together with its humans and spiritual beings) of the Cree people in northeast Canada (Berkes, 1999). So, within (certainly some) Indigenous worldviews, 'culture' and 'nature' are not separate concepts: one's close relations, for example, can include creatures and features of the natural world.

Indigenous knowledge systems are founded in understandings of complexity and relationality—matters that Western scholarship is only just starting to grasp, having been for centuries captured by ideologies that value reductionism and individualism. Key concepts that are shared across many Indigenous societies include relatedness, respect and the importance of reciprocity with the natural world (Artelle et al., 2018). Some Western theories and perspectives in recent decades are starting to articulate similar ideas of relationality, inseparability and the power of non-human entities in human affairs. This includes socio-ecological scholars as cited above, and fields such as new materialism (Grusin, 2015) and ecofeminism (Plumwood, 2005). Donna Haraway (2015) calls for the need to 'make kin' with the non-human world which echoes long-standing Indigenous understandings of the world. Certain philosophical and spiritual belief systems in the Western world similarly recognise our deep interconnections with the natural world, such as the 'deep ecology' movement that had its beginnings in the 1970s (Naess, 2008) and the legacy of St Francis of Assisi in the Christian church. Although culture in English is set up linguistically as an oppositional idea to nature, it doesn't follow that it has to be so conceptually.

Together, these perspectives challenge the conceptual boundaries around the scope of culture. Including nature (or aspects of nature) within an understanding of culture has important implications for sustainability transitions. By offering the understanding that culture can be inclusive of nature, possibilities arise for resetting problematic Western ideologies that underpin the divided way we think about the world and thus how we act in it.

Seeing culture as also inclusive of nature invites questions such as: *Do relational worldviews tend to give rise to more sustainable outcomes than reductive worldviews? What distinctive practices align with concepts of relationality and reciprocity, and what can be learnt from these to assist in sustainability transitions? How can relational knowledge-practice systems be better supported and practitioners empowered?*

What Culture Is Not

At this juncture, I fear that these wide-ranging definitions of culture may have led my readers into a pit of despair. Is culture everything? Is everything culture? If this were the case, the concept would be of no use to anyone. So let me put some boundaries around culture.

Culture is not about how people operate as individuals, each with their unique personal history, genetic inheritance and psychology. It is not about demographics—our age, income, gender and so on, although these may be associated with a tendency to align with particular cultural characteristics. Nor is culture about features that all humans share as social beings—our need for food, shelter, affection and belonging. All of these things may interplay with culture, and their realisation may be culturally shaped, but they are not culture.

Furthermore, culture is never a single phenomenon. Culture is not a piece of clothing or a particular practice; it is always about the interplay between phenomena, such as the meaning of a head covering or the practices associated with a technology. But while singular items are not culture, they may well be cultural. That is, there will likely be cultural influences at play in the design of the piece of clothing; a belief may well be expressed in tangible ways through practices; a routine was learned from others. Accordingly, culture is the opposite of a reductive approach to understanding the social world—it speaks to relationships between people as well as the relationships between the things that people think, have and do.

Finally, culture is not an exclusive explanation. While the social world can be described in cultural terms, it can be described equally validly using other concepts and theories of social, political or psychological processes. My self-appointed task in this book is to elevate culture as a useful lens through which to investigate the causes of and potential solutions to the sustainability crisis, but it is by no means the only approach.

Conclusion

This chapter, I hope, has made it clear why culture can be such a slippery, problematic term. If we use culture as an explanation without defining its meaning, it won't necessarily be obvious to others which sense of the word we intend. Even in our own minds, we may be confused about which of these interpretations of culture we mean. I hope that my broad categorisations, although roughly sketched, are helpful in clarifying culture's divergent interpretations. Each cluster of meanings has a subtle or significant difference to the others in terms of the phenomena it encompasses.

Each approach also has different implications for the roles culture could play in the sustainability transition. Culture-as-nurture invites a more caring approach to processes of production from nature. Culture-as-progress can be redefined as advancing towards more sustainable outcomes through cultural change. Culture-as-product highlights how creative processes and outputs are critically important in conveying and normalising ideas about sustainability. Culture-as-lifeways can help reveal how particular ways of living give rise to sustainable or unsustainable outcomes. Culture-as-meaning invites a focus on the symbolism inherent in products and practices. Culture-as-structure alerts us to pervasive and enduring ideologies and social arrangements that can be barriers to transition. Culture-as-practice invites a focus on routine behaviours and their implications for change processes. Culture-as-purpose highlights how intentional cultural change can achieve more sustainable outcomes. Culture-as-nature pushes past the nature/culture dichotomy to position the natural world as kin and therefore part of a given culture.

Each of the nine conceptualisations of culture is underpinned by extensive literature specific to a particular group of scholars. Each approach has an extraordinarily rich knowledge base that could actively contribute to the analysis and action required for the sustainability transition.

But culture's diverse interpretations bring their own problems. They seem so disparate, and each so conceptually distinct, that we seem no further advanced with my quest for accessibility and integration. The metaphor of blind men seeking to understand an elephant is apt: each exploring a different part of its body and concluding that this is what the animal comprises, without an overall appreciation of the whole. In the following chapter, I move on from describing ears and trunk and feet and tail, and seek to identify the fundamental characteristics of elephantness. Instead of identifying differences, I seek similarities across these interpretations of culture.

References

Adams, R., Martin, S., & Boom, K. (2018). University culture and sustainability: Designing and implementing an enabling framework. *Journal of Cleaner Production, 171*, 434–445.

Adger, W. N., Brown, K., & Conway, D. (2010). Progress in global environmental change. *Global Environmental Change, 20*(4), 547–549.

Adorno, T. W., & Horkheimer, M. (1944). Dialectic of enlightenment (1997th ed.). Verso.

Almila, A.-M. (2016). Actor network theory and its cultural uses. In D. Inglis & A.-M. Almila (Eds.), *The sage handbook of cultural sociology* (pp. 131–143). Sage.

Alvesson, M., & Sveningsson, S. (2015). *Changing organizational culture: Cultural change work in progress*. Routledge.

Artelle, K. A., Stephenson, J., Bragg, C., Housty, J. A., Housty, W. G., Kawharu, M., & Turner, N. J. (2018). Values-led management: The guidance of place-based values in environmental relationships of the past, present, and future. *Ecology and Society, 23*(3).

Bennett, T., Grossberg, L., & Morris, M. (Eds.). (2005). *New keywords: A Revised vocabulary of culture and society*. Blackwell.

Berkes, F. (1999). *Sacred ecology: Traditional ecological knowledge and resource management*. Taylor & Francis.

Berkes, F., Colding, J., & Folke, C. (Eds.). (2003). *Navigating social-ecological systems: Building resilience for complexity and change*. Cambridge University Press.

Bernandi, F., Gonzalez, J. J., & Requena, M. (2006). The sociology of social structure. In B. Bryant & D. Peck (Eds.), *21st century sociology: A reference handbook* (pp. 162–170.). Sage.

Bourdieu, P. (1977). *Outline of a theory of practice*. Cambridge University Press.

Bourdieu, P. (1984). *Distinction: A social critique of the judgement of taste*. Harvard University Press.

Bourdieu, P. (1990). *The logic of practice*. Stanford University Press.

Bridges, K. W., & McClatchey, W. C. (2009). Living on the margin: Ethnoecological insights from Marshall Islanders at Rongelap atoll. *Global Environmental Change, 19*(2), 140–146.

Browne, K. (2005). *An introduction to sociology*. Polity Press.

Burns, E. A. (2020). Thinking sociologically about regenerative agriculture. *New Zealand Sociology, 35*(2), 189–213.

Chaney, D. (1994). *The cultural turn: Scene setting essays on contemporary cultural history*. Routledge.

Collini, S. (Ed.). (1993). *Arnold: "Culture and anarchy" and other writings*. Cambridge University Press.

Comyns, B. (2018). Climate change reporting and multinational companies: Insights from institutional theory and international business. *Accounting Forum, 42*(1), 65–77.

Corsini, F., Laurenti, R., Meinherz, F., Appio, F. P., & Mora, L. (2019). The advent of practice theories in research on sustainable consumption: Past, current and future directions of the field. *Sustainability, 11*(2), 341.

Daddi, T., Bleischwitz, R., Todaro, N. M., Gusmerotti, N. M., & De Giacomo, M. R. (2020). The influence of institutional pressures on climate mitigation and adaptation strategies. *Journal of Cleaner Production, 244*, 118879.

del Reyes-Santiago, M., & R., Sánchez-Medina, P. S., & Díaz-Pichardo, R. (2017). Eco-innovation and organizational culture in the hotel industry. *International Journal of Hospitality Management, 65*, 71–80.
Duit, A., Galaz, V., Eckerberg, K., & Ebbesson, J. (2010). Governance, complexity, and resilience. *Global Environmental Change, 20*(3), 363–368.
During, S. (Ed.). (1999). *The cultural studies reader*. Psychology Press.
Escobar, A. (1996). Construction nature: Elements for a post-structuralist political ecology. *Futures, 28*(4), 325–343.
Faulkner, S. L., Baldwin, J. R., Lindsley, S. L., & Hecht, M. L. (2006). Layers of meaning: an analysis of definitions of culture. In J. R. Baldwin, S. L. Faulkner, M. L. Hecht, & S. L. Lindsley (Eds.), *Redefining culture: Perspectives across the disciplines*. Lawrence Erlbaum Associates.
Ferguson, N. (2012). *Civilisation*. Penguin.
Flamholtz, E. G., & Randle, Y. (2014). Implications of organizational life cycles for corporate culture and climate. In B. Schneider & K. M. Barbera (Eds.), *The oxford handbook of organizational climate and culture* (pp. 235–256). Oxford University Press.
Folke, C. (2006). Resilience: The emergence of a perspective for social-ecological systems analyses. *Global Environmental Change, 16*(3), 253–267.
Foucault, M. (1978). *Power/Knowledge*. Pantheon.
Foucault, M. (1980). *The History of Sexuality*. Vintage.
Galpin, T., Whittington, J. L., & Bell, G. (2015). Is your sustainability strategy sustainable? Creating a culture of sustainability. *Corporate Governance, 15*(1), 1–17.
Geertz, C. (1973). *The interpretation of cultures: Selected essays*. Basic Books.
Giddens, A. (1979). *Central problems in social theory*. Macmillan Press.
Giddens, A. (1984). *The constitution of society: Outline of the theory of structuration*. University of California Press.
Gosnell, H., Gill, N., & Voyer, M. (2019). Transformational adaptation on the farm: Processes of change and persistence in transitions to 'climate-smart' regenerative agriculture. *Global Environmental Change, 59*, 101965.
Grabherr, G. (2009). Biodiversity in the high ranges of the Alps: Ethnobotanical and climate change perspectives. *Global Environmental Change, 19*(2), 167–172.
Grusin, R. (Ed.). (2015). *The nonhuman turn*. University of Minnesota Press.
Gunderson, L. H., & Holling, C. S. (2002). *Panarchy: Understanding transformations in human and natural systems*. Island Press.
Hammersley, M. (2019). *The concept of culture: A history and reappraisal*. Palgrave Macmillan.
Hannerz, U. (1992). *Cultural complexity: Studies in the social organization of meaning*. Columbia University Press.

Hanquinet, L., & Savage, M. (Eds.). (2015). *Routledge international handbook of the sociology of art and culture*. Routledge.
Haraway, D. (2015). Anthropocene, Capitalocene, Plantationocene, Chthulucene: Making Kin. *Environmental Humanities, 6*, 159–165.
Hays, S. (1994). Structure and Agency and the Sticky Problem of Culture. *American Sociological Association, 12*(1), 57–72.
Hays, S. (2000). Constructing the Centrality of Culture—and Deconstructing Sociology? *Contemporary Sociology, 29*(4), 594–602.
Heath, M. T. P., & Chatzidakis, A. (2012). 'Blame it on marketing': Consumers' views on unsustainable consumption. *International Journal of Consumer Studies, 36*, 656–667.
Ingold, T. (2003). Two reflections of ecological knowledge. In G. Sanga & G. Ortalli (Eds.), *Nature knowledge: Ethnoscience, cognition and utility* (pp. 301–311). Berghan Books.
IPES-Food. (2018). *Breaking away from industrial food and farming systems: Seven case studies of agroecological transition*. Food and Agriculture Organisation.
Jasanoff, S., & Kim, S.-H. (2015). *Dreamscapes of Modernity: Sociotechnical Imaginaries and the Fabrication of Power*. University of Chicago Press.
Kasanga, L. A. (2015). Semiotic landscape, code choice and exclusion. In R. Rubdy & S. B. Said (Eds.), *Conflict, exclusions and dissent in the linguistic landscape* (pp. 123–144). Palgrave Macmillan.
Koohafkan, P., & Altieri, M. A. (2016). *Forgotten agricultural heritage: Reconnecting food systems and sustainable development*. Routledge.
Kroeber, A. L., & Kluckhohn, C. (1952). Culture: A critical review of concepts and definitions. *Papers, Peabody Museum of Archaeology and Ethnology, 47*(1). Harvard University.
Lentz, C. (2017). Culture: The making, unmaking and remaking of an anthropological concept. *Zeitschrift Fur Ethnologie, 142*(2), 181–204.
Linnenluecke, M. K., & Griffiths, A. (2010). Corporate sustainability and organizational culture. *Journal of World Business, 45*(4), 357–366.
Maffi, L. (2005). Linguistic, cultural and biological diversity. *Annual Review of Anthropology, 34*, 599–617.
Matinaro, V., & Liu, Y. (2017). Towards increased innovativeness and sustainability through organizational culture: A case study of a Finnish construction business. *Journal of Cleaner Production, 142*, 3184–3193.
Mead, H. M. (2016). *Tikanga Maori: Living by Maori values*. Huia Publishers.
Naess, A. (2008). The shallow and the deep, long-range ecology movement. A summary. In R. Attfield (Ed.), *The ethics of the environment* (pp. 115–120). Routledge.

Obal, M., Morgan, T., & Joseph, G. (2020). Integrating sustainability into new product development: The role of organizational leadership and culture. *Journal of Small Business Strategy, 30*(1), 43–57.

Ortner, S. B. (1984). Theory in anthropology since the sixties. *Comparative Studies in Society and History, 26*(1), 126.

Parsons, T. (1951). *The social system*. The Free Press.

Patterson, O. (2014). Making sense of culture. *Annual Review of Sociology, 40*, 1–30.

Patterson, O., Mohr, J., & Friedland, R. (2004). Culture and continuity: Causal structures in socio-cultural persistence. In R. Friedland & J. Mohr (Eds.), *Matters of culture: Cultural sociology in practice* (pp. 71–109). Cambridge University Press.

Plumwood, V. (2005). *Environmental culture: The ecological crisis of reason*. Routledge.

Pretty, J. (2011). Interdisciplinary progress in approaches to address social-ecological and ecocultural systems. *Environmental Conservation, 38*(2), 127–139.

Rapport, D. J., & Maffi, L. (2011). Eco-cultural health, global health, and sustainability. *Ecological Research, 26*(6), 1039–1049.

Reckwitz, A. (2002). Toward a theory of social practices: A development in culturalist theorizing. *European Journal of Social Theory, 5*(2), 243–263.

Røpke, I. (2009). Theories of practice: New inspiration for ecological economic studies on consumption. *Ecological Economics, 68*(10), 2490–2497.

Schatzki, T. R. (1997). Practices and Actions: A Wittgensteinian Critique of Bourdieu and Giddens. *Philosophy of the Social Sciences, 27*(3), 283–308. 1

Schatzki, T. R., Cetina, K. K., & von Savigny, E. (Eds.). (2001). *The practice turn in contemporary theory*. Routledge.

Scott, J. (2007). Giddens and cultural analysis. In T. Edwards (Ed.), *Cultural theory: Classical and contemporary positions* (pp. 83–107). Sage.

Sewell, W. H. (1992). A theory of structure: Duality, agency, and transformation. *American Journal of Sociology, 98*(1), 1–29.

Shove, E. (2010). Beyond the ABC: Climate change policy and theories of social change. *Environment and Planning A, 42*(6), 1273–1285.

Shove, E. (2014). Putting practice into policy: Reconfiguring questions of consumption and climate change. *Contemporary Social Science, 9*(4), 415–429.

Shove, E., & Pantzar, M. (2007). Recruitment and reproduction: The careers and carriers of digital photography and floorball. *Human Affairs, 17*, 154–167.

Shove, E., Pantzar, M., & Watson, M. (2012). *The dynamics of social practice: Everyday life and how it changes*. Sage.

Stoddard, I., Anderson, K., Capstick, S., Carton, W., Depledge, J., Facer, K., & Kartha, S. (2021). Three decades of climate mitigation: Why haven't we bent the global emissions curve? *Annual Review of Environment and Resources, 46*, 1–37.

Strengers, Y., & Maller, C. (Eds.). (2014). *Social practices, intervention and sustainability: Beyond behaviour change*. Routledge.

Tylor, E. B. (1920). *Primitive culture*. J.P. Putnam's Sons.

United Nations Environment Programme, & and Climate and Clean Air Coalition. (2021). *Global methane assessment: Benefits and costs of mitigating methane emissions*. UNEP.

Welch, D., & Yates, L. (2018). The practices of collective action: Practice theory, sustainability transitions and social change. *Journal for the Theory of Social Behaviour, 48*(3), 288–305.

Williams, R. (1980). *Problems in materialism and culture: Selected essays*. Verso.

Williams, R. (1976). *Keywords: A vocabulary of culture and society*. Oxford University Press.

Wilson, A. M. (2001). Understanding organisational culture and the implications for corporate marketing. *European Journal of Marketing, 35*(3–4), 353–367.

Wuthnow, R., Hunter, J. D., Bergesen, A. J., & Kurzwell, E. (1984). *Cultural analysis: The work of Peter L. Berger, Mary Douglas, Michel Foucault, and Jurgen Habermas*. Taylor & Francis.

Yunkaporta, T. (2020). *Sand talk: How indigenous thinking can save the world*. HarperOne.

CHAPTER 3

Culture's Convergence

Introduction

If we are serious about resolving the sustainability crisis, then it is essential to understand how culture operates as both an impediment and a solution. But to achieve this, culture must first become more accessible. It needs to be a concept that community members, businesses and policy agencies can grasp and apply to their own contexts, and which academics and researchers from any discipline can use as an analytical lens. To return to the analogy at the end of Chapter 2, culture must be understood in a way that reflects the whole elephant while still recognising its parts. We need to be able to be clear when we are referring to a particular interpretation of culture, and how this relates to other interpretations. It should be possible to share insights between different theoretical perspectives using a common language. Finally, culture needs to be conceptualised in a way that does not automatically exclude non-Western worldviews, and instead is open and inviting to all knowledge systems.

What if, instead of saying 'culture is many different things', we took the starting position that 'culture is something and there are many different ways of looking at it'? What would the 'something' of culture then look like? Does culture have a core set of qualities that are shared by all or most of the interpretations (the features of the 'elephant'), or is it academically unsound to suggest that there might be similar features across diverse ontologies and epistemologies?

J. Stephenson, *Culture and Sustainability*,
https://doi.org/10.1007/978-3-031-25515-1_3

Certainly, each of the nine approaches to culture identified in Chapter 2 reflects a different concept of what culture is, how it is constituted, and what is considered to be reliable evidence about culture. For example, culture-as-product focuses on aesthetic (and other) qualities of a circumscribed group of products of human endeavour; culture-as-meaning concerns itself with interpreting the meanings and symbolism that are conveyed by objects, texts, discourses and actions; while culture-as-nature sees humans as relationally entangled with the natural world. On the face of it, these perspectives seem irreconcilable, as they appear to reflect entirely different phenomena. Many academics would consider it heresy to attempt to draw any integrative lines between them.

I disagree. I believe it is entirely appropriate to seek conceptual commonalities across theoretical approaches even though the approaches themselves may represent different ontologies and epistemologies. In this I am heartened by Elinor Ostrom's perspective that research in complex fields may require the use of several theories to explore the patterns of relationships between elements, and that, over and above theories, it is useful to identify a general set of variables that can be used to frame and delimit a field of inquiry (Ostrom, 2005).

My intent in this chapter is, therefore, to do the opposite of the previous chapter. Instead of looking at the differences between approaches to culture, I seek qualities of culture that are common to all or most of the interpretations. Drawing from definitions of culture sourced from all nine approaches, I identify features and qualities of culture that are repeatedly referenced. From this, I attempt to sketch out, at a high level, certain shared characteristics of the entire elephant. From this, I propose a vocabulary to support an integrative understanding of culture.

Seeking Common Qualities of Culture

There are literally hundreds if not thousands of definitions of culture and from time to time brave academics have set themselves to review them. In 1952, Kroeber and Kluckhohn reviewed around 180 academic definitions. They identified six main types: descriptive definitions (enumerations of the content of culture); definitions with an emphasis on social heritage or tradition; normative definitions (relating to rules, values, ideals, behaviour); psychological definitions (learning, habit, problem-solving); structural definitions (referring the pattern or organisation of culture); and definitions relating to the generation of products, artefacts, ideas or symbols (Kroeber & Kluckhohn, 1952). Just over half a century later, Faulkner et al. (2006) reviewed around 300 definitions (until their method reached saturation) and came up with seven clusters of definitions: those that emphasised culture as structure (a system of elements); culture as function (a tool for delivering ends); as a process (e.g. ongoing social construction); as products (artefacts with or without deliberate symbolic intent); as individual or group cultivation of refinement; as power or ideology of a group; and as membership relating to a place or group.

I drew from these two reviews and from other definitions sourced from the literature that informs Chapter 2 to identify common characteristics of culture. From the definitions I was seeking two things: descriptions of the features of culture, and descriptions of culture's qualities, by which I mean its processes or dynamics. Below I discuss representative examples of definitions that relate to each interpretation of culture. My purpose here is not to provide an exhaustive list—that has already been done by others—but to illustrate the range of cultural qualities encompassed by the nine approaches to culture, and to highlight recurring concepts.

Culture-as-nurture, as in agriculture, is defined in the Cambridge English Dictionary as 'to breed and keep particular living things in order to get the substances they produce' or more generally 'the tending *of* something' (emphasis in original) (Williams, 1976: 77). Here, at an abstract level, culture is set of practices that involve maintaining living things and ultimately producing valued products.

Culture-as-progress is defined as 'a general process of intellectual, spiritual and aesthetic development' or 'a standard of perfection' measured either by material progress or by a set of 'higher standards' (Bennett et al., 2005: 80). In this sense culture is a process of change as well as an idealised or sought-after outcome.

Culture-as-product refers to 'the works and practices of intellectual and especially artistic activity' (Williams, 1976: 80); or in more detail, 'the sum of all intellectual and artistic works and processes, primarily within literature, visual arts, music, theatre, films and other artforms: genres used for making and using a special type of human artefacts that rose above vernacular usefulness and reached out into a symbolic sphere of imagination and ideality' (Fornas, 2017: 35). Other definitions have extended culture-as-product to include such things as urban form, architecture, archaeological sites and cultural landscapes (e.g. UNESCO,

1972). With culture-as-product, culture is a specific type of practice, product or place where the symbolic value or meaning is more prominent than its functional value.

An early definition of culture-as-lifeways was already quoted in Chapter 2: 'that complex whole which includes knowledge, belief, art, law, morals, custom, and any other capabilities and habits acquired by man [sic] as a member of society' (Tylor, 1920: 1). A 1940 definition includes physical manifestations of culture: 'an organized body of conventional understandings manifest in art and artefacts which, persisting through tradition, characterizes a human group' (cited in Kroeber & Kluckhohn, 1952: 61). In a recent review of anthropological approaches to culture, it is described as 'the set of learned routines (and/or their material and immaterial products) that are characteristic of a delineated group of people' (Brumann, 1999: 6). Consistently across these definitions, culture is a coherent and interactive system involving beliefs, understandings, products, rules and routines that are characteristic of a particular group of people.

The preamble to the Universal Declaration on Cultural Diversity (UNESCO, 2001) includes aspects of both culture-as-product and culture-as-lifeways, referring to it as '...the set of distinctive spiritual, material, intellectual and emotional features of society or a social group, and [...] encompasses, in addition to art and literature, lifestyles, ways of living together, value systems, traditions and beliefs'.

An early exponent of culture-as-meaning defined it as 'an historically transmitted pattern of meanings embodied in symbols, a system of inherited conceptions expressed in symbolic forms by means of which men [sic] communicate, perpetuate and develop their knowledge about and attitudes towards life' (Geertz, 1973: 89). Other definitions in this conceptual tradition include 'the symbolic-expressive aspect of human behaviour' (Wuthnow et al., 1984: 3) and 'sets of beliefs or values that give meaning to ways of life and produce (and are reproduced through) material and symbolic forms' (Crang, 1998: 2). Here, culture comprises shared meanings and symbolism that are often conveyed by texts, discourses, practices and artefacts.

Some definitions span culture-as-lifeways and culture-as-meaning, such as this from sociologist Sharon Hays: 'Culture encompasses language, symbols, rituals, everyday practices, values, norms, ideas, the categories of thought and knowledge, and the material products, institutional practices, and ways of life established by these' (Hays, 2000: 597). Another definition by a group of sociologists in cultural studies proposes that culture encompasses '(1) ideas, knowledge (correct, wrong, or unverifiable belief), and recipes for doing things; (2) humanly fabricated tools (such as shovels, sewing machines, cameras and computers); and (3) the products of social action that may be drawn upon in the further conduct of social life (a dish of curry, a television set, a photograph, or a high-speed train, for example)' (Hall et al., 2003: 7). Cultural features in these hybrid definitions include tangible elements such as practices, tools and material products as well as symbols, values, ideas, norms, together with shared language and forms of knowledge.

In definitions associated with culture-as-structure, structure is often described in terms of the institutions that shape social life at macro, meso and micro levels, but its fundamental elements are described as the 'norms, beliefs, and values that regulate social action' (Bernandi et al., 2006: 163). Sociologist William Sewell describes structure as comprising a number of dynamically interacting qualities which together form 'cultural schemas' (Sewell, 1992: 11). These include the rules of social life such as 'various conventions, recipes, scenarios, principles of action, and habits of speech and gesture' (Sewell, 1992: 8); human resources such as skills, knowledge, discourses, routines and behavioural patterns; and physical resources that confer value and social power. At a macro-scale, these may form institutions, which in a sociological sense mean established systems of these elements, especially where they acquire a tangible and nameable form, such as marriage, democracy or patriarchy. Here, culture's qualities include beliefs, conventions, habits, knowledge and skills as well as material items, and the ways in which these form dynamic, powerful and enduring systems.

In the field of culture-as-practice, a practice is 'a routinized way in which bodies are moved, objects are handled, subjects are treated, things are described and the world is understood' (Reckwitz, 2002: 250). The most widely-cited definition, already quoted in Chapter 2, is 'a routinized type of behaviour which consists of several elements, interconnected to one other: forms of bodily activities, forms of mental activities, "things" and their use, a background knowledge in the form of understanding, know-how, states of emotion and motivational knowledge' (Reckwitz, 2002: 249). Practice is simultaneously the conjunction of these elements forming an entity, and a performance that carries this conjunction of elements into the future (Shove et al., 2012). Culture here involves bodily routines, ways of thinking, doing and knowing, and how these interact with material items to form enduring patterns.

For culture-as-purpose, where the focus is intentional cultural change, definitions typically include norms, rules and rituals; shared values, ideologies and beliefs; and shared understandings and meanings (Linnenluecke & Griffiths, 2010). Some refer also to behaviours, as in 'the values, beliefs, and norms or behavioural practices that emerge in an organization' (Flamholtz & Randle, 2014: 247).

For definitions of culture-as-nature, I turn to Māori knowledge, where the closest equivalent term is 'tikanga'. Hirini Moko Mead describes tikanga as an interrelated complex of concepts, beliefs, knowledge, practices, rituals and obligations (Mead, 2016). While these elements are not dissimilar to previous definitions of culture-as-lifeways, a key difference is that the underpinning concepts do not locate humans outside of nature, but as part of a constellation of familial relationships that extend to non-human entities. Codes of behaviour, beliefs and knowledge all incorporate aspects of the known natural world as relations for whom one has reciprocal responsibilities and obligations. Culture is, therefore, a relational concept that extends to include the natural world as a cultural actor.

The Commonly Identified Elements of Culture

The first impression from the previous section may be of a bewildering array of concepts. But on closer examination, clusters of similar features and qualities are evident. In this section, I highlight commonly expressed features of culture, and in the following section, I discuss frequently identified cultural processes or dynamics.

Across the definitions of culture that I reviewed, of which a few are quoted above, I found that the following elements are repeatedly identified:

- Ways that people think, including shared and learned meanings, beliefs, understandings and forms of expression
- What people do, including performances, routines and behaviours
- What people have, including products that they make, objects that they acquire, and things with which they interact

I will discuss each in turn.

Ways That People Think

The definitions of culture use a range of terms for phenomena that shape how people think about and understand their world.

Norms: Many of the definitions refer to shared concepts that guide people to think or do things in certain ways that are considered 'correct' for that culture group. Terms used to describe these concepts included laws, rules, norms, conventions, obligations, customs, traditions and recipes. I have selected 'norms' as an overarching term for this group of attributes as it is a generic term for shared prescriptions about what people should do in a given situation. Generally, if norms are flouted, there is some expectation of disapproval or punishment from the wider cultural group (Patterson, 2014). Norms operate at multiple scales: for example they may be shared by a small group (e.g. a dress code) or may be shared at supra-national scales (e.g. norms relating to democratic processes). Norms may become formalised rules (e.g. laws, regulations) but generally are informal but nonetheless strongly influential.

Values: Some of the definitions use terms like values, value systems, principles and morals to refer to shared concepts about what is important to a given culture group. These are higher-level guides for correct or desirable actions when situations are not covered by norms or rules.

Beliefs: A number of definitions of culture refer to even more abstract sets of mental processes, using terms such as beliefs, ideas, categories of thought and concepts. These generally refer to mental representations that underpin how people perceive and understand the world, including shared understandings about what is true or real.

Knowledge: Most definitions refer to aspects of knowledge, either generically (knowledge, forms of knowledge, cultural knowledge) or specifically (e.g. technical knowledge, institutional knowledge, background knowledge). Knowledge in this sense does not have to meet any validity test—some definitions refer to common sense and conventional understandings. Some definitions also refer to embodied knowledge such as skills, know-how, capabilities and bodily techniques.

Symbolism: Terms in this cluster include symbols, meanings, symbolic-meaningful systems, symbolic devices and symbolic-expressive aspects of human behaviour. Here, definitions are referring to how abstract meanings (e.g. values, beliefs, norms) are represented and 'carried' by particular shared understandings, behaviours, objects or forms of communication.

Definitions within culture-as-meaning tend to focus solely or predominantly on these less tangible qualities of culture. Culture-as-product is interested in the significance and symbolism of particular classes of human creations and practices. Definitions in most other interpretations of culture include at least some of the qualities described above as well as more tangible elements. Shared norms, values, beliefs, knowledge, language and symbolism are thus commonly understood features in all approaches to culture although more implicit in culture-as-nurture.

What People Do

Although we generally tend to think of culture as exemplified by habitual actions and routines, any human action can be culturally influenced. So while some words and phrases in the definitions relating to 'doing' refer specifically to routines, others refer to actions more generally.

Routines: Many definitions included terms that referred to regularly repeated activities, including habits, routines, practices, everyday practices, rituals, traditions, habits of speech and gesture, behavioural patterns and habitual behaviours. A major part of social existence comprises actions or activities that occur repeatedly involving similar actions and sequences. These include such things as daily domestic routines, rituals such as religious ceremonies, or standardised social interactions such as the way formal meetings operate. Many routines are learned in childhood and carried through one's life, while others are learnt in specific settings such as school or work.

Actions: Some definitions do not differentiate between repeated and rare behaviours, and so by implication include all actions that people undertake. Such terms in the definitions include behaviour, bodily activities, intellectual and artistic practices, and the making of things. These terms reflect that the whole of human behaviour, whether habitual or not, can be both shaped by and conveys shared forms of cognition. For example, activities that might take place only once or twice in a lifetime such as buying a house are strongly influenced by cultural norms, values and beliefs. People may also act in non-routine ways when faced with changing circumstances, and these actions may be strongly culturally influenced or may disrupt routines. While habitual actions are important in the replication of culture, occasional actions are no less cultural and may be critical for action on sustainability issues.

In culture-as-nurture, activities include breeding, keeping and tending. Culture-as-product recognises especially skilled practices with symbolic value. For culture-as-lifeways, actions and practices are seen as intimately tied with how people think, and are often dependent on learned bodily routines. For culture-as-meaning, the focus is on the imputed meaning and symbolism of actions. Routines and behavioural patterns at larger scales are integral to culture-as-structure. Culture-as-practice focuses on routines as part of the bundle of ideas that comprises practice, while culture-as-purpose is interested in changing behaviours along with ways of thinking.

What People Have, Use and Make

In this category are things that are made, acquired and used, regardless of scale and purpose. Terms used in the definitions include artefacts, products, art, intellectual and artistic works, tools, infrastructure, land, technology and things. These may range in scale from objects that are made or owned by individuals to large-scale resources such as urban form, and may include things that have virtually no physical presence such as software.

Such items play different roles in interpretations of culture. In culture-as-product, the focus is on things that are intentionally produced to represent and convey meanings and ideas. In culture-as-meaning, the interest is in what things symbolise. Objects are considered as part of culture in their own right in culture-as-lifeways, as part of cultural assemblages. In culture-as-practice, things are an integral part of routines. In culture-as-structure, material items concretise ideologies and institutions. In culture-as-nature, aspects of the natural world are kin and thus part of one's culture.

The features that appear repeatedly across definitions of culture can thus be clustered into three distinct elements: ways in which people think about and understand the world; what they do both routinely and occasionally; and material items that they have, use and make. Some of the nine perspectives on culture put greater or complete emphasis on just one or two of these elements, but even there, the other elements are rarely absent. For example, culture-as-meaning focuses on beliefs and symbolism but elicits these meanings from studies of what people do and have, including means of communication such as text. Culture-as-product focuses on works of art (things) and performance (activities) in relation to the beliefs, values and symbolism that they convey. Most definitions also pay particular attention to ways in which these elements interact, a topic discussed in the next section.

Table 3.1 summarises this discussion. Common cultural features across definitions can be clustered into three (interactive) elements: how people think (norms, values, beliefs, symbolism, and cognitive and bodily knowledge); what people do (routines and actions); and material items (products and acquisitions).

Culture's Dynamic Qualities

Culture is more than just some or all of these elements—in most definitions it is also a dynamic process and/or a distinctive and enduring set of relationships between people. Anthropologist Marvin Harris describes culture as 'the total socially acquired life-way or life-style of a group of people. It consists of the patterned, repetitive ways of thinking, feeling, and acting that are characteristic of the members of a particular society or segment of a society' (Harris, 1975: 144). Sharon Hays describes it as 'a social, durable, layered pattern of cognitive and normative systems at once material and ideal, objective and subjective, embodied in artefacts and in behavior, passed about in interaction, internalized in personalities, and externalized institutions. [...] Culture is both the product of human interaction and producer of certain forms of human interaction' (Hays, 1994: 65). For sociologist Orlando Patterson, it is 'a dynamically stable process of collectively made, reproduced, and unevenly shared knowledge structures that are informational and meaningful, internally embodied, and externally represented and that provide predictability, coordination equilibria, continuity, and meaning in human actions and interactions' (Patterson, 2014: 1). From these descriptions, and drawing also from other definitions as discussed earlier in this chapter, the characteristics that stand out as particularly relevant to our interest in sustainability are cultural membership, cultural learning, culture's systemic qualities and culture's durability.

Table 3.1 Widely shared features in definitions of culture, clustered into three cultural elements

Cultural elements	*Terms often used in definitions*	*Summary term*
How people think	Laws, rules, norms, conventions, obligations, customs, traditions	Norms
	Values, value systems, principles of action, morals	Values
	Beliefs, ideas, categories of thought, concepts	Beliefs
	Knowledge, forms of knowledge, cultural knowledge, technical knowledge, institutional knowledge, background knowledge, motivational knowledge, common sense, conventional understandings	Cognitive knowledge
	Skills, know-how, capabilities, bodily techniques	Bodily knowledge
	Symbols, meanings, symbolic devices, symbolic-expressive aspects of human behaviour	Symbolism
What people do	Habits, routines, practices, everyday practices, rituals, traditions, habitual behaviours	Routines
	Behaviour, bodily activities, behavioural patterns, intellectual and artistic practices, the making and use of things	Actions
What people have, use and make	Things, artefacts, products, art, intellectual and artistic works, tools, infrastructure, technology	Products and acquisitions

Cultural Membership

The concept of culture is generally associated with membership: the idea that a roughly definable group of people share some cultural elements, such as meanings, objects, practices and/or forms of communication. Similar patterns of cultural elements are adhered to (loosely or strictly) within the group. Terms in the definitions above that reflect this idea of membership include 'human group' and 'social group'.

A culture group can be almost any scale larger than a single individual and smaller than all of humankind. At one extreme, a culture group could comprise a single family with unique aspects to their patterns of interaction and communication. At the other extreme, Western culture with its tendency towards individualism and nuclear family structure is widely (but not universally) shared among people living in Western nations. When we speak of culture groups, they may be at any scale within this continuum.

Culture groups are not necessarily clearly bounded, and cultural membership is rarely static. New people may enter the group and adopt cultural features, while others may drop out, or membership may be fleeting. People's degree of adherence to cultural elements may vary greatly, and group members may show a 'greater or less degree' of sharing (Brumann, 1999: 4). Importantly, culture is not always consciously adopted. It can be so embedded that people do not even recognise that they adhere to cultural norms until they are faced with another cultural experience. Those members who adhere (more or less) to a particular group of cultural elements may change over time. The boundaries of membership may also alter if previously important cultural elements become less important to the group. Cultural membership is thus fuzzy and dynamic.

Individuals can belong to more than one culture group. Many people are accustomed to moving between one set of cultural expectations in their family life, another at work, and maybe others in religious, recreational or educational settings. In this sense, they share some cultural elements—beliefs, practices, etc.—with others in one setting, and share another set of cultural elements in a different setting. As individuals we may thus have membership in several cultures with which we align (consciously or otherwise), but this does not mean that we have an individual culture. Culture is always about something collective.

Membership is a core feature of most interpretations of culture, although not to culture-as-nurture and only tangentially to culture-as-product. Within many Indigenous perspectives, members of one's culture group may well include features of the natural and/or sacred world with whom one is kin or shares obligations, such as animals, natural features and spirit beings (Harmsworth & Awatere, 2013; Turner & Spalding, 2013; Yunkaporta, 2020).

Cultural Learning

Culture is learned from and reinforced by other people, a process that starts in childhood and continues through life. People develop 'clusters of common concepts, emotions and practices' as a result of regular interactions with others (Brumann, 1999: 1). In the modern era we are influenced by other cultural vectors such as advertising, social media and social influencers. Language, including bodily signs, group-specific jargon, dialects, speech patterns and intonations, is often distinctive to a culture group and core to both cultural identity and cultural communication. Social interactions and other forms of messaging convey rules and expectations about what to do, say, think or have in order to be acceptable to the wider culture group. If a culture group is small, some or all of the members may engage in face-to-face interactions, but for larger groups norms and rules may be absorbed from media, entertainment and online interactions, so that even where individuals don't personally know others in the group, they still adopt particular cultural elements. I discuss processes of cultural learning more fully in Chapter 4 in the section on cultural vectors.

Culture is continuously under active construction as individuals adjust their actions to take into account the actions and reactions of others (Cipolletta et al., 2020; Coleman, 1986). Cultural expectations can be enforced through a range of measures, from disapproving body language through to the enforcement of institutionalised rules such as laws. Although being a member of a culture group can be an affirming experience, it can also be disempowering and hurtful, both within a culture group (for those who fail to live up to the fullest expression of cultural alignment) and between culture groups (where difference may be used as the basis for dismissing, blaming or attacking others).

Cultural Systems

Many of the approaches to culture recognise its systemic qualities. This is reflected in terms used in some of the definitions above, such as 'complex whole', 'organised body', 'system', 'interrelated complex', 'layered pattern', and 'dynamically stable process'. These convey how a given culture comprises not just certain distinctive elements, but how these interact and align. For example, shared norms and meanings may strongly influence what objects people aspire to acquire; culturally aligned skills and knowledge may determine the routines that people undertake. These system-like interactions tend to reinforce particular arrangements of cultural elements, here called cultural ensembles, and thus the reproduction of similar cultural patterns over time.

Culture-as-lifeways, culture-as-structure and culture-as-practice take a particular interest in how this system, involving the dynamics within cultural ensembles, maintains cultural stability. Cultural reproduction is a key concept in culture-as-structure, with structuration being particularly focused on how the routinised actions of people, following shared 'rules', reproduce those rules. Culture-as practice focuses on how specific practices or clusters of practices are sustained over time through continuous reproduction. In culture-as-meaning, texts and discourses are carriers of cultural meaning and thus part of the process of cultural reproduction. In culture-as-product, artefacts and performances consolidate, create and communicate cultural ideas.

Cultural Endurance

These qualities of culture mean that it has a tendency to endure, with relatively constant members and relatively similar cultural ensembles. This durability arises from the ways in which humans find affirmation in belonging; the ways in which culture is learned and passed on between group members; the existence of shared rules and expectations and the perceived repercussions of flouting these; and the systemic interactions between cultural elements such that they tend to reinforce each other. This combination means that cultural ensembles and memberships can remain relatively stable over long periods of time. This is one of the reasons why culture is such a challenge for the sustainability transition, but also thankfully a reason why some more sustainable cultures have endured. Chapter 5 elaborates on processes of cultural stability using research-based examples.

Despite this, cultures can and do change. Cultural membership can swell or shrink; cultural actors can adopt new ways of thinking, acting and possessing; and changes to context or to specific cultural elements can cause cascading change to cultural ensembles. Processes of cultural change (in the context of sustainability) are the focus of Chapter 6.

Table 3.2 Widely shared concepts in definitions of culture: cultural qualities

Cultural qualities	*Terms used in definitions*
Cultural membership	Delineated group of people, features of a society or social group, members of particular society or segment of a society, features of the natural and/or sacred world with whom one is kin
Cultural learning	Learned routines and concepts, historically transmitted, socially acquired, transferred through language, signs and experiences
Cultural systems	Symbolic-meaningful systems, a system of inherited conceptions, elements interconnected to one another, complex whole, ways of life, organised body, cultural schema, institution, produced and reproduced through material and symbolic forms
Cultural endurance	A dynamically stable process, continuity, predictability

Table 3.2 summarises the cultural qualities that have commonly appeared across the nine different approaches to culture. While not all concepts appear in all definitions, and the emphasis varies according to definition, they are sufficiently shared to form the basis of an integrative understanding of culture.

Conclusion

Three main conclusions can be drawn from this high-level analysis of convergences across approaches to culture.

First, despite its different interpretations, there is a surprising commonality regarding culture's features and qualities. This was not my expectation when I began the review. Three clusters of features emerged: ways of thinking, ways of doing and material items, which I describe as the core elements of culture. Across definitions, these are generally understood to be dynamically linked such that they shape and are shaped by each other. Most interpretations see culture as specific to a group of people and learned and passed on over time. For all of these reasons, culture has a tendency to endure in a similar form over time.

Although these features and qualities of culture are not all expressed all approaches to culture, they are sufficiently universally shared that any of the approaches would be able to 'see' aspects of itself represented. They describe qualities of the whole 'elephant' of culture, offering an integrative set of concepts which are inclusive of its different interpretations.

All approaches to culture have the potential to usefully support sustainability endeavours, as I discussed in Chapter 2. At present, however, these possibilities are largely invisible and unavailable unless one is embedded in the relevant discipline or sub-discipline. Given the importance of culture as both a constraint and a pathway to a more sustainable future, my goal is to make it more widely accessible as a framing for thinking, analysis and action. From this perspective, the features and qualities described in this chapter show promise for developing an analytic vocabulary. In the next chapter, I discuss how these ideas have helped with some modifications to the scope and vocabulary of the cultures framework so that it encompasses the range of cultural features and qualities identified here.

A further insight from this review has been that most approaches to culture have been more interested in its durability than in its ability to change. For the most part, explanations of culture are interested in how and why it remains relatively stable over time. Culture-as-meaning, culture-as-lifeways, culture-as-structure and culture-as-practice all bring scholarly insights into why and how culture acts as a constraint against change. From a sustainability perspective, it is certainly important to be able to understand how and why it is so hard to change culture. But it may be even more critical to understand how and why cultures change and how to stimulate cultural change towards more sustainable outcomes. Although some academic work has used cultural theory to illuminate opportunities for change (e.g. Shove & Spurling, 2013), this is far from common.

The cultures framework was originally designed to explore processes of cultural change and appears to help fill a gap in conceptualising and analysing change processes from micro to macro scales. In the next chapter, I introduce the cultures framework, and in subsequent chapters, I describe examples of its application to explore culture's transformability as well as its durability.

References

Bennett, T., Grossberg, L., & Morris, M. (Eds.). (2005). *New keywords: A revised vocabulary of culture and society*. Blackwell.

Bernandi, F., Gonzalez, J. J., & Requena, M. (2006). The sociology of social structure. In B. Bryant & D. Peck (Eds.), *21st century sociology: A reference handbook* (pp. 162–170). Sage.

Brumann, C. (1999, February). Writing for culture: Why a successful concept should not be discarded. *Current Anthropology, 40*, 1–17.

Cipolletta, S., Mascolo, M. F., & Procter, H. (2020). Intersubjectivity, joint action and sociality. *Journal of Constructivist Psychology, 35*(3), 904–929.

Coleman, J. S. (1986). Social theory, social research, and a theory of action. *American Journal of Sociology, 91*(6), 1309–1335.

Crang, M. (1998). *Cultural geography*. Routledge.

Faulkner, S. L., Baldwin, J. R., Lindsley, S. L., & Hecht, M. L. (2006). Layers of meaning: An analysis of definitions of culture. In J. R. Baldwin, S. L. Faulkner, M. L. Hecht, & S. L. Lindsley (Eds.), *Redefining culture: Perspectives across the disciplines*. Lawrence Erlbaum Associates.

Flamholtz, E. G., & Randle, Y. (2014). Implications of organizational life cycles for corporate culture and climate. In B. Schneider & K. M. Barbera (Eds.), *The Oxford handbook of organizational climate and culture* (pp. 235–256). Oxford University Press.

Fornas, J. (2017). *Defending culture: Conceptual foundations and contemporary debate*. Palgrave Macmillan.

Geertz, C. (1973). *The interpretation of cultures: Selected essays*. Basic Books.

Hall, J. R., Neitz, M. J., & Battani, M. (2003). *Sociology on culture*. Routledge.

Harmsworth, G., & Awatere, S. (2013). Indigenous Māori knowledge and perspectives of ecosystems. In J. R. Dymond (Ed.), *Ecosystem services in New Zealand: Conditions and trends* (pp. 274–286). Manaaki Whenua Press.

Harris, M. (1975). *Culture, people, nature: An introduction to general anthropology* (2nd ed.). Crowell.

Hays, S. (1994). Structure and agency and the sticky problem of culture. *American Sociological Association, 12*(1), 57–72.

Hays, S. (2000). Constructing the centrality of culture—And deconstructing sociology? *Contemporary Sociology, 29*(4), 594–602.

Kroeber, A. L., & Kluckhohn, C. (1952). Culture: A critical review of concepts and definitions. *Papers, Peabody Museum of Archaeology and Ethnology, 47*(1). Harvard University.

Linnenluecke, M. K., & Griffiths, A. (2010). Corporate sustainability and organizational culture. *Journal of World Business, 45*(4), 357–366.

Mead, H. M. (2016). *Tikanga Maori: Living by Maori values*. Huia Publishers.

Ostrom, E. (2005). *Understanding institutional diversity*. Princeton University Press.

Patterson, O. (2014). Making sense of culture. *Annual Review of Sociology, 40*, 1–30.

Reckwitz, A. (2002). Toward a theory of social practices: A development in culturalist theorizing. *European Journal of Social Theory, 5*(2), 243–263.

Sewell, W. H. (1992). A theory of structure: Duality, agency, and transformation. *American Journal of Sociology, 98*(1), 1–29.

Shove, E., Pantzar, M., & Watson, M. (2012). *The dynamics of social practice: Everyday life and how it changes*. Sage.
Shove, E., & Spurling, N. (Eds.). (2013). *Sustainable practices: Social theory and climate change*. Routledge.
Turner, N., & Spalding, P. R. (2013). "We might go back to this": Drawing on the past to meet the future in northwestern North American indigenous communities. *Ecology and Society, 18*(4), 29.
Tylor, E. B. (1920). *Primitive culture*. J.P. Putnam's Sons.
UNESCO. (1972). *International convention for the protection of the world's cultural and natural heritage*. UNESCO.
UNESCO. (2001). *Universal declaration on cultural diversity*. UNESCO.
Williams, R. (1976). *Keywords: A vocabulary of culture and society*. Oxford University Press.
Wuthnow, R., Hunter, J. D., Bergesen, A. J., & Kurzwell, E. (1984). *Cultural analysis: The work of Peter L. Berger, Mary Douglas, Michel Foucault, and Jurgen Habermas*. Taylor & Francis.
Yunkaporta, T. (2020). *Sand talk: How indigenous thinking can save the world*. HarperOne.

CHAPTER 4

The Cultures Framework

Introduction

In Chapters 2 and 3 I focused on the question: 'what is culture?' From this chapter onwards, my focus turns to how to depict culture's relationship with sustainability-related outcomes. This requires both a narrower and a wider framing than the discussion in the previous two chapters. Narrower, because it means paying attention to particular aspects of actors' cultures that have a causal relationship with sustainability outcomes. Wider, because although cultural theories can help in this quest, other bodies of knowledge also contribute.

As briefly outlined in Chapter 1, the cultures framework was initially developed to underpin research on the poor uptake of energy efficiency by households and businesses. Its development was informed by a range of theories, only some of which could be called cultural. The framework has since been widely used to underpin research on sustainability-related questions and has proved its usefulness as a sound and flexible analytical structure. Over time, the framework has changed subtly as its core language was adjusted to reflect its broadening scope beyond energy topics. In this chapter, I introduce a new iteration of the framework, drawing from my work in developing this book—from both my review of cultural theories (Chapters 2 and 3) and my review of its many applications around the world (Chapters 5 and 6). Readers interested in the backstory of the cultures framework are encouraged to read earlier papers

J. Stephenson, *Culture and Sustainability*,
https://doi.org/10.1007/978-3-031-25515-1_4

that track its applications and evolution over time (Stephenson, 2018; Stephenson et al., 2010, 2015a).

The cultures framework offers an analytical structure to describe and investigate culture's relationship with sustainability. It is centred on actors, by which I mean that it is designed to explore the cultural characteristics of individuals, households, communities, businesses, organisations, sectors or other collectives at any scale. It provides a scaffold for identifying cultural characteristics of the relevant actors that give rise to sustainability-related outcomes. In addition, the framework draws attention to actors' agency, and to broader influences that shape culture and its consequences for sustainability. While the cultures framework represents a widely shared set of concepts about culture, it extends beyond culture to depict the interplay between culture, the agency of cultural actors and their context.

I start this chapter by explaining what I mean by framework, differentiating between theories and frameworks. Because the cultures framework seeks to represent the causal relationships between culture and sustainability outcomes, I then discuss how the concept of causality can be applied to complex systems such as culture. For the remainder of the chapter, I present the cultures framework in stages, first explaining each of its component parts and how these have conceptually evolved, and then how it operates as a complete framework.

Theories and Frameworks

Theories are abstract explanations or conceptualisations of the workings of some aspect of the cosmos. They specify the scope of an inquiry, offer propositions and testable hypotheses, provide a descriptive vocabulary and are usually specific to a discipline or field of study. Whether explicitly or not, theories assume an ontology (the aspect of existence that is 'real' for the purposes of the study) and epistemology (the forms of knowledge that can describe that 'reality'). For example, theories of chemistry are founded in an ontology of material substances, classical physics is interested in forces and fields, and psychology focuses on the processes of the human mind. Theories thus shape what features or qualities of a phenomenon are studied and how they are studied. They can be powerful and effective for in-depth investigations of a particular aspect of the world. However, they can also constrain thinking on complex issues if they force a narrow focus and reductive consideration of the phenomenon being studied.

Traditionally, the scientific method has largely relied upon single-theory investigations, arguing that combining ontologies and epistemologies is bad science. This may be the case where a discipline has particular expectations of what constitutes evidence and proof. However, it is increasingly evident that this does not necessarily produce the best results where the problem under investigation is complex and multi-faceted (Sabatier & Weible, 2014). Relying on single theories can create cognitive presuppositions whereby only parts of complex situations and processes are recognised, and important aspects may be missed. Transformative agendas such as sustainability transitions require collaboration-inducing frameworks that can represent and include diverse forms of knowledge and multiple ways of understanding the world (Wyborn et al., 2019). Using several theories allows complex problems to be investigated from different perspectives, avoiding 'capture' by the assumptions that are built into theories, and (if done well) enabling robust conclusions supported by multiple sources of evidence.

Frameworks are a way of helping to organise a multi-theoretical approach. The term 'framework' is sometimes equated with 'theory' by social scientists, but I adopt Elinor Ostrom's perspective that frameworks operate at a meta-theoretical level (Ostrom, 2005). Frameworks set out highly generalised variables and indicate the relationships between them. By depicting a complex field through certain universal qualities and dynamics, a framework can help researchers to formulate questions and thence identify the theories that are best suited to answering those questions. Once analysis has been undertaken, a framework can help with integrating the findings to describe the field from multiple perspectives.

The cultures framework was initially developed as a framework in Ostrom's sense, to depict a set of high-level variables that were of shared relevance to a multidisciplinary team of researchers. Across many studies, and at more generalised levels, it has continued to work effectively as a framework that supports multi-theory, multidisciplinary research. At the same time, the cultures framework can be used as a theory in its own right, albeit one that is open to a wide range of methods of inquiry. In this latter respect, it has underpinned many studies as a theory about how cultural dynamics give rise to sustainability outcomes. This ability to use it as either a framework or a theory will be further elaborated in Chapter 8, where I discuss its use to underpin research.

Culture and Causality

Almost any question to do with cultural processes will involve highly complex sets of variables, and especially so where we are needing to consider the interplay between social and physical realms, as with sustainability issues. I find it helpful to think of culture as a complex adaptive system, a term applied to open, dynamic, self-organising systems that involve exchanges and transformations of information, energy and other resources (Turner & Baker, 2019). Self-organisation refers to the concept that no single part of the system has direct or exclusive control over the system as a whole; instead the system's dynamics are the result of multiple ongoing physical and/or cognitive interactions. External forces—those over which these systems have little or no direct control—can shape the system because it has the ability to respond and adapt. Complex systems thus involve countless interactions between countless phenomena, including feedback loops that can stabilise or amplify features of the system. Common features of complex adaptive systems include path dependence, temporality, non-linearity, emergence and adaptive capacity, all of which are features of culture. Theories of complexity are thus helpful in understanding the dynamics of cultural processes.

We are accustomed to thinking about causality as a linear process: action X gives rise to Y which gives rise to Z. In most scientific studies, statistical evidence is necessary to prove cause-and-effect relationships between variables, but understanding causal processes in complex social systems is a very different consideration. Here causality is the result of multiple diverse factors interacting over time which may be ontologically and epistemologically distinct (e.g. symbolism, human actions, physical technologies, power relationships, policies). Causal reconstructions that seek to explain outcomes need to decide which variables and processes to focus on. This involves identifying high-level and influential processes among variables that can be observed playing out over multiple cases. Causal claims in complex social systems are, therefore, rarely able to be statistically proven because there are too many variables and/or they cannot all be measured. Instead claims of causality must rely on repeated observable and traceable mechanisms.

These mechanisms—the means by which outcomes are brought about—can be explored at a general level or confined to a narrow range of variables and processes. General mechanisms refer to 'general activities and social interactions that bring about change' (Geels, 2022:

10). These include actors and their properties, and the activities they engage in by themselves or with others. Assumptions regarding these properties and mechanisms will differ according to the choice of theories. An interpretivist analysis, for example, would focus on the role of sense-making and meaning, while conflict theory would focus on power differentials and struggles, and actor-network theory would be interested in the agentive power of technologies. Specific causal mechanisms are less abstract and 'provide causal logics for more focused topics and issues' (Geels, 2022: 10). Examples relevant to sustainability questions include the theory of adoption of new innovations (Rogers, 2003), theory of planned behaviour (Ajzen, 1991) and theories of socio-technical transitions (Smith, 2007). Conceptual frameworks that seek to show how complex systems operate generally accommodate multiple theoretical models at both general and specific levels that elucidate different causal mechanisms and indicate their relationships.

In a complex system, although no single part can control the whole, a change in one variable can have considerable influence on other parts of the system and thus on the outcomes in which we are interested. In addition to the systems' tendency to self-organisation, it is necessary to pay attention to processes of systemic change, including how and where change is initiated. This means we need to account for agency—the ability of actors in the system to make choices that have repercussions for other parts of the system. Examples from a sustainability perspective could include a decision by an investor to remove their funds from a fossil-fuel intensive sector, an angler's decision to fish in a protected area, or a commuter's decision to walk to work rather than drive. Agency adds further complexity to analysing systems, because the social world cannot be analysed as if it were a machine with fully predictable relationships between its parts: humans, thankfully, can do the unexpected.

In many respects, then, culture operates as a complex adaptive system. Its sustainability outcomes rarely involve linear causality, but generalised and specific mechanisms can be identified. Culture has a tendency towards self-organisation but is also open to change, both from the agentive power of cultural actors and from broader influences on culture.

Of course, culture is not the only contributor to sustainability outcomes. In the case of climate change, for example, causal explanations of human failure to take action include (amongst numerous other explanations) individuals' psychology (Gifford, 2011), market failures

(Stern, 2007) and vested interests (Franta, 2021). The cultures framework brings forward the role of culture as another (often overlooked) causal contributor which is a complex system in its own right. The sustainability outcomes of this system depend on the features and dynamics of the cultural ensemble in question and also, importantly, the choices of individual actors. Together, these interactions have implications for social, economic and environmental measures of sustainability.

Overview of the Cultures Framework

The cultures framework is a set of interrelated high-level concepts that represent a complex adaptive system with culture at the core. These concepts and the relationships are visually represented as a model (Fig. 1.2 in Chapter 1)—a cognitive aid offering a 'thinkable format for reflection on and reasoning about the domain in question' (Harré, 2009: 133). Models use a set of concepts that attempt to capture the essential characteristics of a field, and a simplified logic to infer relationships between the parts. Each concept represents a distillation of an aspect of the field of study and is given a representative term or phrase.

Words matter, particularly when one is seeking to communicate ideas precisely. As should be clear from Chapter 2, words can have multiple meanings, and may mean different things to different people. A word in one setting can have a very different set of connotations in another, even where it appears to refer to the same thing. Certain words also carry the weight of their own academic histories, having been chosen to represent a specific academic concept, often particular to a discipline or theory. Choosing the right terms to represent a set of ideas can be fraught.

With the cultures framework, the choice of words to describe the core set of stripped-down concepts has evolved over time. This is partly a result of the expanding scope of the model from energy topics ('energy cultures framework') to the more generic 'cultures framework'. Additionally, some of the core terms have been adjusted to ensure they represent the right scope of meaning, are reasonably straightforward to understand at face value, and are as free from theoretical confusion as possible.

The following sections introduce the cultures framework in stages. I first introduce the core elements of the cultural ensemble ('motivators', 'activities' and 'materiality') and their dynamics. I then discuss the means by which culture is learned, enabled and communicated ('vectors'). This framing might be sufficient to study culture itself as a field of enquiry, but the cultures framework has a broader purpose: to hypothesise the causal role of culture in relation to sustainability. To support this particular agenda, the framework incorporates three further core concepts; the first relating to constraints on cultural actors' abilities to make independent choices (the 'agency barrier); the second representing the context that shapes a given culture ('external influences'); and the third reflecting the implications of the interactions of all of these for selected measures of sustainability ('sustainability outcomes').

The diagrams in this section, and indeed throughout the book, are purposefully drawn with basic online drawing tools so that anyone can replicate them with ease. This aligns with my aim to democratise culture as a concept; to make it readily understandable and useable as an analytical approach by anyone in any situation.

The Cultural Ensemble

The term 'cultural ensemble' describes the dynamic whole of the three core elements of culture—how people think, what they do, and associated material items. Ensemble is a term most commonly used to refer to a group of musicians, actors or dancers who perform together, but is also applied to a group of any items viewed as a whole. I have adopted 'ensemble' in both of these senses: a group of features viewed as a whole that are also dynamically interactive. In the cultures framework, the cultural ensemble is shown as three core elements linked by two-way arrows (Fig. 4.1). These elements comprise motivators, activities and materiality, and align with the three sets of cultural elements identified in Table 3.1.

Motivators

I use 'motivators' to refer to shared characteristics that shape or influence actors' actions and choices. These include norms, values, beliefs, symbolism, language and cognitive and bodily knowledge: the range of cultural qualities associated with 'how people think' in Table 3.1.

The important roles that these play in thought processes and thereby in actions and decision-making are extensively discussed in psychological, sociological, anthropological and cultural literature. Learned from others and shared with others, they constrain and shape actors' thoughts, judgements, decisions and actions. Cultural motivators exclude psychological

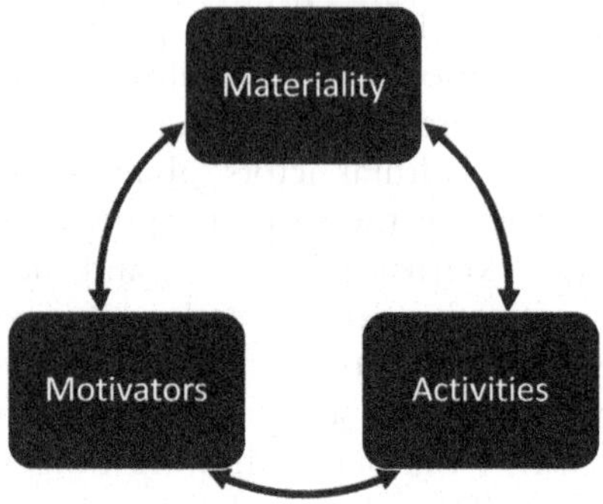

Fig. 4.1 The cultural ensemble

characteristics that are unique to individuals, and those that are shared by humanity generally.

Previous versions of the cultures framework used the term 'norms' rather than 'motivators'. In early research using the cultures framework, shared norms were shown to be a powerful influence on actors' sustainability characteristics (Stephenson et al., 2015a). Norms reflect what actors consider to be 'normal' in their daily lives, establish expectations about how to behave in a particular context and convey the potential for social disapproval if these expectations are not met. Some of our work suggested that it was also important to pay attention to aspirational norms, as these may act as a springboard for change under the right circumstances (Ford et al., 2017). Although norms can be sometimes hard to extract from research participants as they are so bound up in actors' perspectives of normalcy, our team and other researchers have repeatedly found them to be an important cultural influence on sustainability outcomes.

Over time it became clear that 'norms' did not capture a sufficiently wide range of influential cognitive characteristics. Research using the cultures framework showed that beliefs, understandings, meanings, values and forms of knowledge were also important influences on sustainability outcomes. Their role in culture was further emphasised by the review discussed in Chapter 3. Clearly, it was necessary to adopt a term with a broader reach. In the absence of a commonly used umbrella term for 'beliefs, norms, aspirations, meanings, values and forms of knowledge and understanding', I have adopted the term 'motivators'. This term emphasises how shared cognitive characteristics shape and underpin actors' actions and decisions, and fits with the focus of the cultures framework on the relationships between culture and sustainability outcomes.

Activities

'Activities' encompasses 'what people do' in Table 3.1. It replaces the term 'practices', which was used in earlier versions of the cultures framework. Practices had been intended to be interpreted in its everyday meaning of customary actions, and to include both everyday routines and infrequent actions. This is consistent with some academic interpretations (e.g. anthropologist Sherry Ortner calls practice 'all forms of human action' or 'anything people do' [Ortner, 1984: 149]). For sociologists, however, practice tends to be more narrowly interpreted as habitual activities, and in social practice theory, it has an even more specialised interpretation (see culture-as-practice, Chapter 2). This caused confusion among some users of the framework.

'Activities' has less academic load and is generally understood as the things people do. Research using the cultures framework has repeatedly shown how sustainability outcomes can be strongly influenced by one-off or occasional actions (e.g. purchasing a house or car) as well as by everyday routines (e.g. heating practices). Chapter 3's review of definitions reinforces that the 'doing' of culture includes routines (e.g. 'habits', 'rituals', 'habitual behaviours') and less regular actions (e.g. 'behaviour', 'bodily activities', 'actions') (see Table 3.1) refers to all of the 'doings' of cultural actors and includes the full spectrum from regularly repeated routines to occasional or rare actions.

Materiality

It isn't easy to find a word that is sufficiently broad to capture the range of things that people have, make and acquire. 'Material culture' is often used in anthropology to refer to the physical evidence of culture, including tools, objects and structures (Woodward, 2007) and was initially used in the energy cultures framework. The concept includes the understanding that such items have both functional and symbolic qualities, and that people's decisions to make, use, acquire or discard them are significantly shaped by their alignments (or misalignments) with other cultural characteristics, such as beliefs, norms and practices. The shift in terminology to 'materiality' in more recent versions of the framework was largely triggered by the confusion caused by the doubling up of the word 'culture' (as in 'cultures framework' and 'material culture') and also by the particular association of material culture with the discipline of anthropology.

'Materiality' avoids this disciplinary capture and is intended as an umbrella term for physical items, features and products as well as those that scarcely have corporeal form, such as digital phenomena. Its scope aligns with the descriptors in the bottom row of Table 3.1. As with all terms, it is not perfect as it carries an alternate sense of significant or important, but this does not seem to have caused confusion in research using the cultures framework to date.

Cultural Dynamics

Motivators, activities and materiality form the core elements of the cultural ensemble, but they cannot be neatly teased apart; in fact, a large part of what we recognise as culture is the utter entanglement of the mental, physical and active aspects of social life. In this section, I discuss how the cultures framework invites consideration of the dynamics within and between cultural elements, and also cultural dynamics between actors within a culture group. Inconsistencies or reinforcements within or between cultural elements may help explain how and why cultures change or remain relatively static over time.

Cultural dynamics in any given situation will be immensely complex, so as with any systems model the attempt here is to highlight particular interactions that may be useful to consider when undertaking cultural analyses, without constraining consideration of other cultural dynamics.

Dynamics Between Cultural Elements

The interplay *between* cultural elements is indicated by the curved arrows between the three elements (Fig. 4.1). The arrows draw attention to how cultural features are often closely entangled, such as the symbolism implicit in cars, or how social norms shape our clothing choices. The arrows also indicate how cultural features may shape or influence one another, such as how technologies invite particular practices (e.g. smart phones have led to new forms of communication). This doesn't mean that these interplays are always supportive—for example, people may believe in climate change but fail to adopt low-carbon travel practices—but it invites the exploration of any relevant relationships between motivators, activities and materiality.

Inter-element dynamics such as these have often been explored in research using the cultures framework, and they appear to play an important role in both cultural stasis and cultural change. Where cultural features actively support each other, they can effectively create balancing loops (using system dynamics parlance), and the resulting cultural ensemble can be very resistant to change. In other situations, where these linkages are less strong, cultures can change rapidly as a result of a disturbance to one element. I will discuss examples of both situations in Chapters 5 and 6, respectively.

Dynamics Within Cultural Elements

The interplay *within* cultural elements—that is, between the features within an element—can also affect sustainability outcomes (Fig. 4.2). In relation to materiality, for example, if people generate their own electricity through solar power, they may be more likely to want to acquire an electric vehicle. In relation to activities, patterns of church attendance may have implications for cooking routines. Motivators can also interplay, such that safety concerns might override environmental concerns in parents' choices to drive their children to school. Research using the cultures framework has revealed some of these dynamics and examples are discussed in later chapters.

This concept of intra-element dynamics is not visually represented in the main cultures framework, but it is indicated here as circular arrows representing interactions between features within each element.

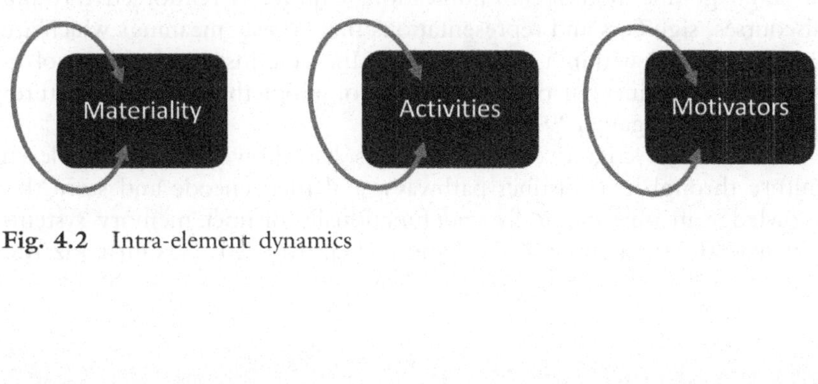

Fig. 4.2 Intra-element dynamics

Dynamics Between Cultural Actors

While we can analyse culture at the level of a single actor, culture is a collective phenomenon. Actors (individuals, households, organisations, etc.) learn culture from others and share their culture with others. This is often an unconscious process but may also be intentional, such as within organisations seeking to develop a particular culture. Interactions between members of a culture group are thus another cultural dynamic to consider: the ways in which cultural features are learned from others, mutually reinforced, and shared with others.

To date, uses of the cultures framework have not generally paid much attention to how culture is learned and transmitted, which is unsurprising as this was never discussed or signalled in seminal papers. However, in a sustainability context, it may be critical to pay attention to these processes. To understand how culture is maintained and replicated, or how it alters over time, it is important to appreciate both the means by which it is learned, and the cognitive and physiological processes that enable culture to be absorbed and acted upon.

Culture is learned and passed on through all of our senses. As children we learn our family's culture through hearing sounds and languages, smelling and tasting foods, seeing and observing what others do, and through bodily interactions such as carrying out tasks and interacting with familiar objects. We learn about the shared significance and meaning of words, objects and practices. At school this learning continues as we are immersed in a new culture, with different languages and signifiers, new bodily skills and new forms of knowledge and understanding. This repeats in different contexts such as being employed in a new workplace or travelling to a different country. We absorb wider cultural concepts through means, such as visual media, social interactions and observations of social life. We model others by learning and adopting particular forms of language and bodily communication. Culture is reinforced through discourses, signifiers and representations that convey meanings which are shared by those within a culture group. In some instances, this involves deliberately closing out others who do not adopt those cultural features (Gray & McGuigan, 1997; Storey, 2018).

Work in the cognitive social sciences has shown that people learn culture through two distinct pathways, and then encode and store this knowledge in physiologically and functionally distinct memory systems that have different retrieval mechanisms (Lizardo, 2017). Omar Lizardo names these as declarative culture and non-declarative culture, but I will use the more easily understood 'semantic' and 'bodily' knowledge. Semantic knowledge is mostly what Lizardo calls 'know-that' and is mainly in the form of propositions about the world that are removed from a personal context. For the most part semantic knowledge is learned through spoken and written language and acquired through a relatively small number of exposures. Examples include lay knowledge absorbed from childhood, formal learning in education and understandings picked up through media. Bodily knowledge is about 'know-how' and is acquired quite differently, being built up from long-term repeated exposure to consistent patterns of experience. It involves the repetition of bodily actions, repetitive use of perceptual and motor skills, and recurring cognitive and emotional messages picked up about the world. This is a 'slow learning' pathway that builds up through habituation and the learning of skills. The existence of two separate routes for the acquisition and retrieval of cultural knowledge has immense implications for cultural replication and cultural change, and for the design of policy interventions.

Through these cultural vectors, people absorb similar routines to others, adopt or make similar material items, and/or develop similar norms or beliefs. The very existence of common motivators, objects and activities can facilitate inter-group dynamics, creating a strong feeling of identity. This sense of being members of a group that shares cultural features may be overt, such as where people actively align (e.g. members of Extinction Rebellion) or membership may be relatively invisible to adherents, especially where shared features are ubiquitous (e.g. the ubiquitous role of automobiles in everyday life in most developed countries).

There may also be important interplays to consider between different cultural vectors. For example, semantic knowledge will not necessarily be consistent with knowledge learned and held in bodily skills, and this may have implications for sustainability. Research on efficient driving, for example, found that while people cognitively understood how to drive efficiently, their learned bodily skills dominated so that their everyday driving was inefficient (Scott & Lawson, 2018).

Understanding inter-actor cultural dynamics and their role in cultural learning and reinforcement is important for sustainability research. If we are seeking to understand cultural inertia, we need to know how cultural patterns are learned and reinforced between actors in ways that make deviation difficult. If we are interested in cultural change, then it is helpful to be aware of how adjustments in culture are passed from one actor to another within a culture group, and possibly spread beyond a culture group to be adopted by others.

I refer to these dynamics collectively as vectors—ways in which cultural is actively learned and then shared with (and adopted by) others (Fig. 4.3). As with intra-element dynamics, cultural vectors are not visually represented in the standard model of the cultures framework (Fig. 1.2) but may be important to consider in analyses using the framework.

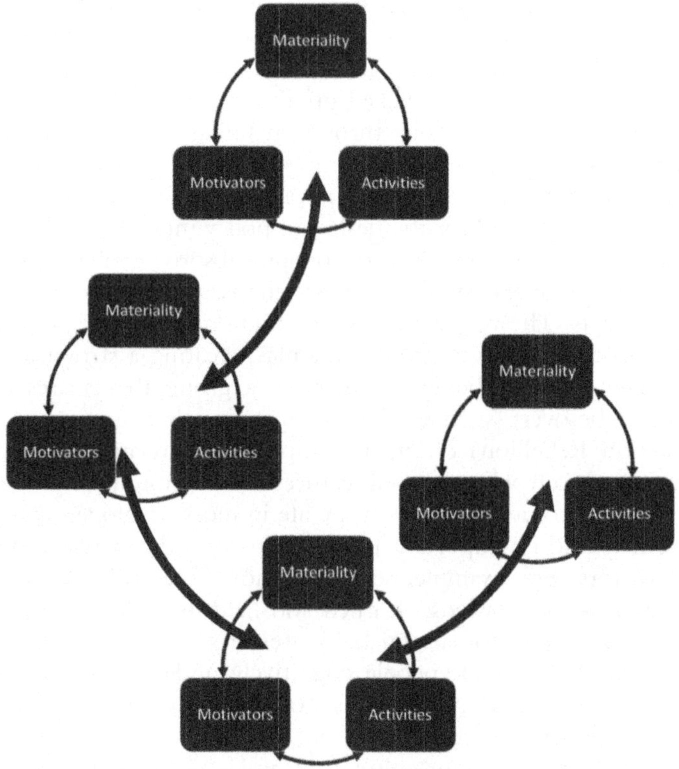

Fig. 4.3 Vectors: processes of cultural learning

Assembling Culture

The conceptual language of the cultures framework describes cultures in terms of certain highly generalised variables. The three core elements of motivators, activities and materiality are strongly aligned with the common elements of culture identified in Table 3.1, Chapter 3. The shared qualities identified in Table 3.2—of cultural membership, cultural learning, cultural systems and cultural endurance—are reflected in inter-element dynamics, intra-element dynamics and cultural vectors.

Weaving these concepts together, I define culture (for the purposes of the cultures framework) as comprising *distinctive patterns of motivators (norms, values, beliefs, knowledge and symbolism), activities (routines and actions) and materiality (products and acquisitions) that form dynamic ensembles which are shared by a group of people and learned through both cognitive and bodily processes.*

This definition draws attention to the three core elements of the cultural ensemble, to the dynamics between those elements and to the processes involved in cultural learning and cultural membership. It invites investigation of culture as a complex dynamic system, as well as more simply as recognisable patterns of elements within a population.

Figure 4.4 is a visual summary of the discussion above, illustrating the concepts that are encompassed by the high-level variables of motivators, activities, materiality and vectors. When this set of concepts is applied in the cultures framework, we purposefully select the motivators, activities and materiality to investigate. The focus is necessarily on those aspects of our actors' cultural ensemble that have a causal relationship with the sustainability outcomes in which we are interested. If we are interested in water consumption, for example, we are unlikely to need to consider actors' commuting routines, but this would be relevant if we were interested in carbon emissions. In the following chapters there are many examples of how other researchers have determined the scope of the motivators, activities and materiality relevant to their field of interest.

Actors, Agency and the Scope of Culture

From this point, I start to introduce additional features of the cultures framework that take it beyond simply a model of culture. I first put bounds around the scope of culture in order to bring into consideration the potential for actors to alter features of their cultural ensembles.

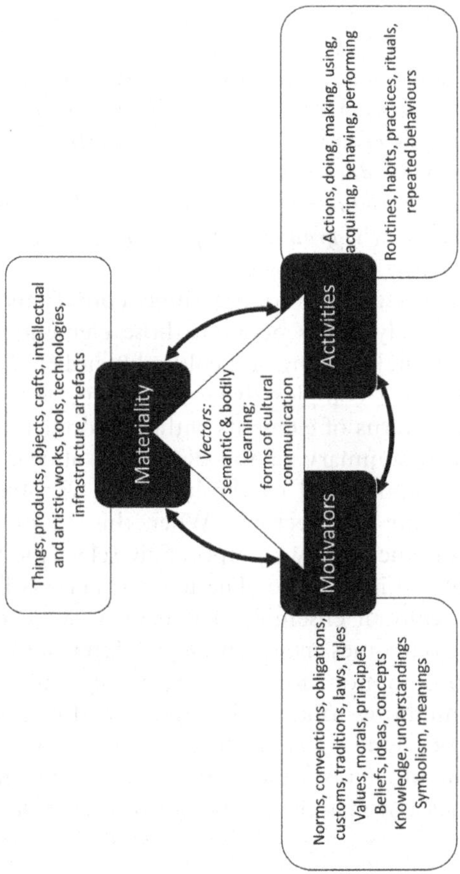

Fig. 4.4 Features of motivators, activities, materiality and vectors

Actors may include individuals, groups, communities households, businesses, organisations or other collectives. The 'unit of analysis' is the cultural ensemble of the actor, with a particular focus on those features and qualities of their ensemble that are causally linked to sustainability outcomes. For example, if we were interested in water pollution from dairy farming, the relevant actors would likely be dairy farmers, and we would seek to understand what aspects of their particular ensemble of motivators, activities and materiality, and the dynamics between these, related to discharges to water. In addition, and critically for sustainability purposes, we would need to consider the extent to which they would be able to make changes to this cultural ensemble if they were so inclined.

Generally, when we think of culture, it is a relatively boundless concept; an aspect of social life that is discernible at every scale from (for example) how individuals prepare food to how the global food system operates. Like any system, culture has no defined edges, and this is one of the reasons why it can be hard to describe and study. For the purposes of the cultures framework, however, we purposefully limit the scope of the aspects of culture that form our core focus. This is a common technique used in systems analysis, where an artificial boundary is placed around the part of the system to be investigated while recognising its interconnections with the wider system. With the cultures framework, this is conceptually achieved by drawing a boundary that distinguishes between the motivators, activities and materiality that the actor potentially has some control over, and those that they do not.

In this way, the concept of agency (the capacity to achieve desired change) is incorporated in the cultures framework. If cultural change is to occur (towards more sustainable outcomes) it will generally involve actor-led adjustments to their motivators, materiality or activities. This requires considerations of the extent to which that actor is able to make deliberate changes. So rather than depicting culture as a relatively fuzzy pattern of features observable in social life, the cultures framework draws a boundary around actors' cultures to represent the point at which their agency diminishes sharply. Beyond this point, cultural characteristics still exist, but for the purposes of the cultures framework these are considered to be external to the actor's cultural ensemble.

The 'agency boundary' is indicated in the cultures framework as a dashed circle around the core dynamics of an actor's cultural ensemblee (Fig. 4.5). Within the circle are the elements and dynamics of the actors' culture, to the extent that they enact it and have the potential to adjust it.

To provide an example of what I mean, I draw from the work of one of my Masters' students who studied the energy cultures of tenants (Nicholas, 2021). She identified material items, norms and practices that tenants had the capacity to alter in order to improve the energy performance of their homes. This formed the extent of their agency, and thereby formed the conceptual boundary around the tenants' energy culture. This was contrasted with the energy culture of their landlords, which included attitudes towards tenants' complaints about damp and cold, their maintenance practices and their ownership of the accommodation and its fixed heating assets. Because of landlord—tenant relationships, there were many actions that tenants could not undertake to improve energy performance and landlords could, but did not necessarily choose to. The agency boundary thus indicates the limits of the actor's capacity to act to change features of their cultures, should they choose to do so. What is 'in' and out' of a cultural actor's agency is always going to be context-dependent, but it has proved to be a fruitful concept in research using the cultures framework.

The agency boundary also draws attention to power differentials. Agency will be influenced by many things. For householders, for example, this might include their financial circumstances, their age or gender, their

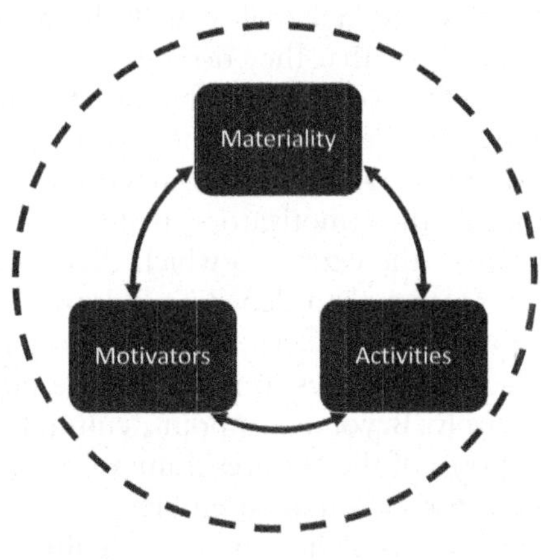

Fig. 4.5 The agency boundary

education, their familiarity with bureaucratic systems, specialist knowledge, or whether they own or rent their home. For actors such as businesses or organisations, agency will be affected by other factors. The greater the agency limitations, the harder it is for actors to adjust their way out of an unsustainable culture, and the more their cultural ensembles will be constrained and shaped by external factors including wider cultural influences.

Agency is a critical consideration for policy or other interventions for change because agency limitations (and related power differentials) can constrain actors' ability to adjust aspects of their cultural ensemble even if they wish to, or prevent them from taking advantage of a policy initiative. I illustrate this point in Chapter 5 with examples, and in Chapters 7 and 8 I show how the concept of agency can be applied in policy development and research.

External Influences

Cultures don't exist as a separate bubble from the rest of the world—they form and evolve in response to their context. They are shaped by history, environmental conditions, political developments, broad ideologies and countless other influences. These can be conceived of as a variety of external or exogenous influences which either support the current cultural ensemble or support cultural change. In organisational theory, the terms transactional and contextual environments are used to make a similar distinction (Emery & Trist, 1965).

The cultures framework places these influences outside the agency boundary, as they are largely beyond any control by individual actors (Fig. 4.6). There are of course myriad contextual factors at play with any culture, but for the purposes of analysis using the cultures framework we limit the range of external influences under consideration. We focus on those that are either supporting cultural stasis and those that are (potentially) tending to drive cultural change. In applying this to personal mobility, for example, investments in motorway infrastructure and carparking tend to support a car-dominant culture while investments in walkways and cycleways may support a shift to more active mobility (Stephenson et al., 2015b). The power of external influences on cultural stasis and cultural change will be discussed further in Chapters 5 and 6.

In Giddensian terms (see culture-as-structure in Chapter 2) the agency boundary can be interpreted as the point of interplay between structure

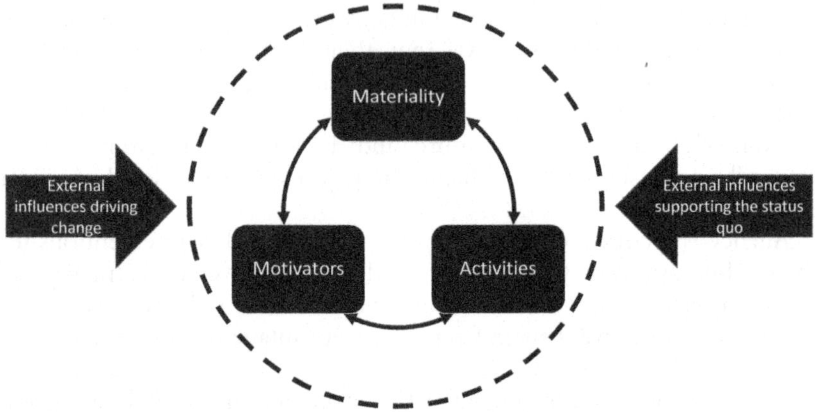

Fig. 4.6 Differentiating between external influences that support cultural stasis and those that support cultural change

and agency, although analysis using the cultures framework is not limited to always considering external influences in structural terms. It is open to other conceptualisations of these broader influences, including envisaging some external influences as being cultural. In other words, the cultures of less powerful actors can be shaped by the cultures of more powerful actors. This is indicated diagrammatically in Fig. 4.7 below. I discuss and illustrate this idea of less powerful cultures being influenced by more powerful ones in Chapter 5.

Although external influences are by definition largely unchangeable by actors, there may be occasions when an actor's influence can reach beyond the agency barrier and help to transform previously unreachable structures and cultures. This transformative potential of culture is discussed further in Chapter 6.

Sustainability Outcomes

The final component of the cultures framework vocabulary is 'outcomes'. Cultural ensembles and their dynamics have consequences for measures of sustainability and indeed for any yardstick (e.g. educational achievement [Hsin & Xie, 2014] or intergenerational inequality [Lareau, 2011]). Outcomes as used here refers to social, economic, environmental or

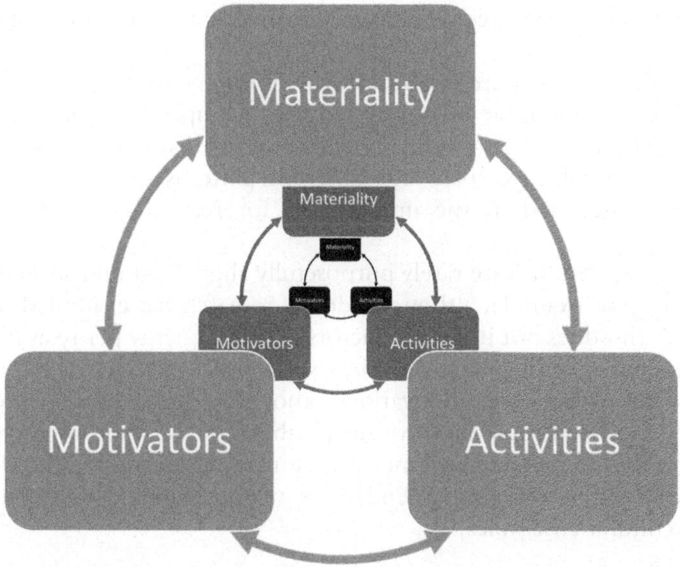

Fig. 4.7 Conceiving of external influences as more powerful cultures

other sustainability-related consequences of actors enacting their cultural ensembles.

The outcomes chosen for study will differ according to the interests of the researcher and the context of the research: outcomes could range from empirical measures of a single dimension such as energy use, to multi-dimensional social, cultural, and spiritual qualities. Research using the cultures framework to date has considered outcomes as varied as gender equity, water consumption, health and wellbeing, and the adoption of renewable energy technologies.

The concept of outcomes is depicted in the framework as an arrow from the heart of the cultural ensemble to a triangle representing sustainability outcomes (Fig. 4.8). This indicates that we are interested in the implications of cultural dynamics (including their interaction with external influences) for the outcomes of interest. The arrow pointing back into the heart of the cultural ensemble is a reminder that achieving change in sustainability outcomes may itself result in consequential change to a cultural ensemble. I refer to the latter as 'proximal outcomes' (i.e. changes to cultural features) and the former as 'distal outcomes' (i.e. changes to sustainability measures). These are illustrated and discussed further in Chapters 7 and 8.

The outcomes that are of interest will determine the pertinent features of the cultural ensemble to be studied. For example, if we are interested in health outcomes of housing retrofits we are likely to be looking at different cultural ensembles and different patterns of occurrence than if we are interested in the implications for reducing greenhouse gas emissions.

Cultural ensembles are rarely purposefully aligned around sustainability outcomes. For example, although a lot of research has examined 'energy cultures', this does not imply that actors perceive energy per se as defining of their identity or the purpose of their social lives. Energy may figure only tangentially within their motivations, knowledge systems, activities and material choices. People may live sustainably (or not) without any particular intention to do so, which means that investigations using the cultures framework often reveal unintended or unconsidered consequences of actors' cultural ensembles.

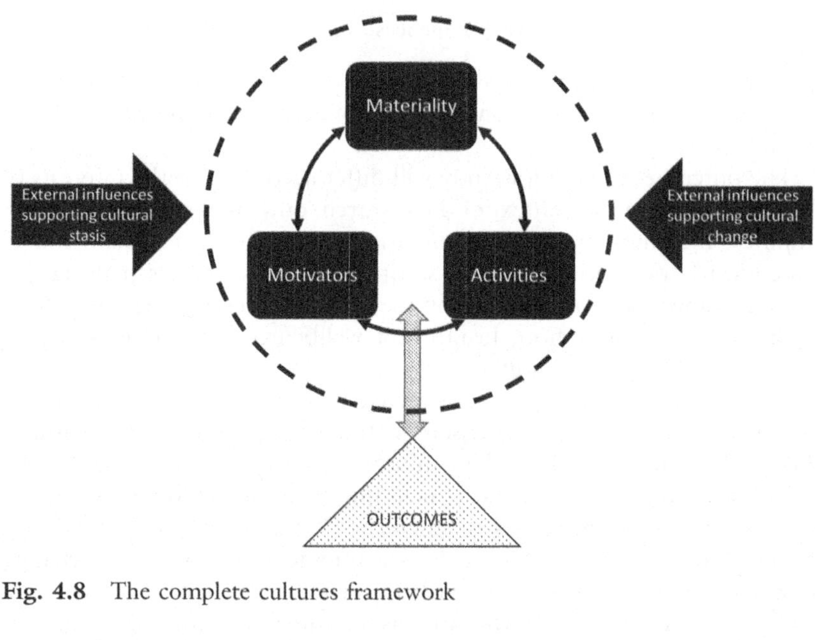

Fig. 4.8 The complete cultures framework

Some actors or groups of actors consciously adopt cultural features that align with their sustainability concerns, such as the upsurge in people becoming vegans or vegetarians, communities seeking to become self-sufficient in renewable energy, and businesses seeking to become carbon zero. In some instances the purposeful adoption by a few actors of new aspirations, beliefs, practices and other shared cultural elements may form the basis of new social movements that can sometimes radically change societies (Snow et al., 2018). Historical examples include anti-slavery movements, union movements, feminism, gay rights and environmental movements. However, in most cases sustainability is only one of multiple concerns of cultural actors, and the links between their cultural ensembles and sustainability outcomes are not necessarily obvious to them. Analysis using the cultures framework can help to reveal how cultural ensembles and their dynamics are complicit in sustainability outcomes.

Conclusion

The cultures framework offers a conceptual structure for visualising and analysing the relationship between culture and sustainability. It is a set of interlinked high-level ideas that bring attention to salient features and dynamics in this field of inquiry. Its language and diagrammatic form reflect well-established understandings about culture and its dynamics, but it is not only a model of culture. Beyond its core concepts of cultural ensembles and their dynamics, the framework adds conceptual elements from other fields of knowledge. From systems theory, it takes understandings of system dynamics and system boundaries. From structuration it adopts the concept of agency and applies it to circumscribe actors' cultures. External influences, which represent contextual pressures on culture, are considerations in many disciplinary fields. Sustainability outcomes refer to how these variables interact and result in tangible sustainability-related consequences.

The diagrammatic form and language of the framework, as described in this chapter, reflect its evolution but are still consistent with its earlier forms. The main changes have been to broaden the scope of some concepts, to add the concept of cultural vectors, and to visually depict 'outcomes' as part of the framework rather than just describing them.

In the next two chapters, I show how the cultures framework has been used to explore numerous sustainability questions in many parts of the world. Most applications to date relate to energy-related topics, which is unsurprising given its origins as the energy cultures framework. Issues explored in this context include energy consumption, energy efficiency behaviours, smart grids, adoption of energy technologies and energy poverty. Actors have included households, age cohorts, businesses, industries, universities, economic sectors and countries. As well as energy topics, it has been widely applied to transport outcomes including mode choices, driving behaviour, freight efficiency and city-wide transport policies. It has also been applied to other sustainability issues such as water, food and climate change. The stories in the next two chapters show how the framework can help reveal the hidden workings of culture in relation to sustainability outcomes.

References

Ajzen, I. (1991). The theory of planned behaviour. *Organisational Behaviour and Human Decision Processes, 50*, 179–211.

Emery, F. E., & Trist, E. L. (1965). The causal texture of organizational environments. *Human Relations, 18*(1), 21–32.

Ford, R., Walton, S., Stephenson, J., Rees, D., Scott, M., King, G., Williams, J., & Wooliscroft, B. (2017). Emerging energy transitions: PV uptake beyond subsidies. *Technological Forecasting and Social Change, 117*, 138–150.

Franta, B. (2021). Early oil industry disinformation on global warming. *Environmental Politics, 30*(4), 663–668.

Geels, F. W. (2022). Causality and explanation in socio-technical transitions research: Mobilising epistemological insights from the wider social sciences. *Research Policy, 51*(6), 104537.

Gifford, R. (2011). The dragons of inaction: Psychological barriers that limit climate change mitigation and adaptation. *American Psychologist, 66*(4), 290–302.

Gray, A., & McGuigan, J. (1997). *Studying culture: An introductory reader*. Bloomsbury Academic.

Harré, R. (2009). Saving critical realism. *Journal for the Theory of Social Behaviour, 39*(2), 129–143.

Hsin, A., & Xie, Y. (2014). Explaining Asian Americans' academic advantage over whites. *Proceedings of the National Academy of Sciences of the United States of America, 111*(23), 8416–8421.

Lareau, A. (2011). *Unequal childhoods: Class, race, and family life*. University of California Press.

Lizardo, O. (2017). Improving cultural analysis: Considering personal culture in its declarative and nondeclarative modes. *American Sociological Review, 82*(1), 88–115.

Nicholas, L. (2021). *On Common Ground: How landlords & tenants shape thermal performance of private rental properties.* Masters thesis, University of Otago.

Ortner, S. B. (1984). Theory in Anthropology since the Sixties. *Comparative Studies in Society and History, 26*(1), 126.

Ostrom, E. (2005). *Understanding institutional diversity.* Princeton University Press.

Rogers, E. M. (2003). *Diffusion of innovations* (5th ed.). Free Press.

Sabatier, P. A., & Weible, C. M. (2014). *Theories of the policy process* (3rd ed.). Westview Press.

Scott, M. G., & Lawson, R. (2018). The road code: encouraging more efficient driving practices in New Zealand. *Energy Efficiency, 11*(7), 1617–1626.

Smith, A. (2007). Translating sustainabilities between green niches and socio-technical regimes. *Technology Analysis & Strategic Management, 19*(4), 427–450.

Snow, D. A., Soule, S. A., Kriesi, H., & McCammon, H. J. (Eds.). (2018). *The Wiley Blackwell companion to social movements.* John Wiley & Sons.

Stephenson, J. (2018). Sustainability cultures and energy research: An actor-centred interpretation of cultural theory. *Energy Research & Social Science, 44*, 242–249.

Stephenson, J., Barton, B., Carrington, G., Gnoth, D., Lawson, R., & Thorsnes, P. (2010). Energy cultures: A framework for understanding energy behaviours. *Energy Policy, 38*(10), 6120–6129.

Stephenson, J., Barton, B., Carrington, G., Doering, A., Ford, R., Hopkins, D., Lawson, R., McCarthy, A., Rees, D., Scott, M., Thorsnes, P., Walton, S., Williams, J., Wooliscroft, B. (2015a). The energy cultures framework: Exploring the role of norms, practices and material culture in shaping energy behaviour in New Zealand. *Energy Research and Social Science, 7*, 117–123.

Stephenson, J., Hopkins, D., & Doering, A. (2015b). Conceptualizing transport transitions: Energy Cultures as an organizing framework. *Wiley Interdisciplinary Reviews: Energy and Environment, 4*(4), 354–364.

Stern, N. (2007). *The Economics of climate change: The stern review.* Cambridge University Press

Storey, J. (2018). *Cultural theory and popular culture: An introduction.* Routledge.

Turner, J. R., & Baker, R. M. (2019). Complexity theory: An overview with potential applications for the social sciences. *Systems, 7*(1), 4.

Woodward, I. (2007). *Understanding material culture.* SAGE Publications.

Wyborn, C., Datta, A., Montana, J., Ryan, M., Leith, P., Chaffin, B., Miller, M., & van Kerkhoff, L. (2019). Co-producing sustainability: Reordering the governance of science, policy, and practice. *Annual Review of Environment and Resources, 44*, 319–346.

CHAPTER 5

Cultural Stability

Introduction

Culture's relative stability is often described as one of its defining features. Certain cultures (or aspects of culture) have long lifetimes, some over centuries or even over millennia. Cultures that change very little can be both beneficial and disastrous for a sustainable future.

There are many lessons to be learned from longstanding cultures that have worked out how to live sustainably, not the least to show that cultures *can* adapt to become sustainable once they become familiar with what it takes to not over-exploit their environment. Apart from a few inspiring examples, this is not something that Western civilisation has yet learnt how to do very well. On the other hand, many more sustainable cultures have suffered greatly from the onslaught of colonialism and modernity but still retain knowledge and worldviews that can play an important role in sustainability transitions. We must honour the tenacity of life-affirming cultures such as these and ensure that they are supported and empowered to act.

But the overwhelming problem for the world is that too many unsustainable cultures are stubbornly resistant to change. We can see these at many scales, from global cultures of unsustainable consumption and production to households and businesses that are locked into unsustainable patterns despite knowing that they should change. To have any chance of transition, we need to know how and why such cultures continue to be replicated. As I have already noted, cultural reproduction

J. Stephenson, *Culture and Sustainability*,
https://doi.org/10.1007/978-3-031-25515-1_5

has been a significant focus of cultural research for many years, particularly theories relating to culture-as-structure and culture-as-practice (Chapter 2). The cultures framework offers a complementary approach to help unpack cultural stability.

In this chapter, I discuss how applications of the cultures framework have helped to reveal processes of habituation, and to show why groups of people maintain similar cultural ensembles even when they aspire to change. I draw from over thirty research-based articles or chapters that have used the cultures framework to examine situations of cultural stasis and/or resistance to change. These papers all use the cultures framework in its earlier forms, so they use the conceptual language of norms, practices and materiality rather than motivators, activities and materiality. The research findings are nonetheless equally relevant and insightful, and indeed some of the findings have contributed to the broadening of the scope of the framework as reflected in the new terminology.

I start the chapter by illustrating ways in which culture plays a role in sustainability-related outcomes. I then discuss examples of consistent cultural patterns that are identifiable across populations. I provide examples of how cultures can resist change through the dynamics within a cultural ensemble. I show how culture can be seen to operate at multiple scales, with broader and more powerful cultures influencing those with less agency, and I finish by discussing examples of external influences shaping culture and constraining change.

Each example I use is unique to the time, place, people and context; what I aim to bring to the fore are generalisations about the workings of culture. The illustrations, sourced from some of these studies, use different iterations of the cultures framework but usefully demonstrate how the framework diagram can be used to summarise key features of cultural ensembles, cultural dynamics and external influences.

Culture and Sustainability Outcomes

It was a visit to Sweden many years ago that first impressed on me the stark differences in energy culture between Scandinavian and New Zealand households. Here, our older houses are traditionally detached, constructed from timber and poorly insulated. Despite recent government initiatives to improve insulation standards and promote insulation retrofitting, most homes are still cold by first-world standards. Many Kiwis know this first-hand from comparing it with their much warmer experiences in centrally heated northern hemisphere homes in winter, and because visitors from Europe and North America often complain about our cold houses.

The most common heating devices in New Zealand homes today are heat pumps and enclosed wood stoves. Generally, these just heat the main living space, although doors might be left open to spread the heat further. Sometimes bedrooms will be heated for a short period before bed time with an electric resistive heater. Centrally heated homes are rare. For most Kiwis it is normal to put on layers of clothes to stay warm when indoors. In our cool bedrooms, we use electric blankets or hot water bottles to warm up in bed. New Zealanders have a particularly frugal 'heating culture'. Having (relatively) cold indoor temperatures is considered normal, and aligns with practices and material choices that have become cemented from childhood. This is not to say that we are all necessarily happy to live in cold homes. As I will show in later examples of energy poverty, people can be forced into an unhealthy heating culture because of circumstances beyond their control.

In contrast, for many readers in similarly cool zones of the global north, your homes will almost certainly have some form of central heating and will be well insulated. Warm inside temperatures throughout your house probably mean that you can wear light clothing indoors most of the year.

Here are two very different cultures related to home heating, with divergent implications for sustainability outcomes. If we were considering the outcomes in relation to, say, the United Nations' Sustainable Development Goal 3 (SDG3) (good health and wellbeing), a greater proportion of New Zealanders may be suffering from the health effects of cold indoor temperatures than Europeans and North Americans. In terms of SDG1 (end poverty), they may be suffering more fuel poverty especially if they attempt to heat to European-equivalent temperatures and homes lack effective insulation (Howden-Chapman et al., 2012).

We can examine different sustainability outcomes from a given culture, but we can also turn this on its head. By starting with a sustainability outcome, we can use the cultures framework to trace the role of culture in influencing that outcome. Using greenhouse gas emissions from transport as an example, we can see how car culture operates at multiple levels or scales: on the one hand creating widely prevalent dependency on a single form of mobility, but on the other hand showing considerable variation in how that dependency plays out across populations. Over the past century, a 'system of automobility' (Urry, 2004) has created our current fossil-dependent mobility—a system that includes the technological dominance of internal combustion engines, powerful corporate actors, global supply chains of fuel provision, governmental investments in roading infrastructure and subsidies supporting fossil fuel use. These and other infrastructures have actively enabled the proliferation of fossil fuel using vehicles that are one of the most problematic sources of greenhouse gas emissions today. This system has normalised individual car ownership to such an extent that, for many, it is hard to imagine life without a car. The prevailing cultural ensemble in most developed countries comprises private automobiles, habitual personal car use even for short trips, low use of public transport or active travel, and social norms that associate cars with autonomy, freedom and perceptions of status (Stephenson et al., 2015; Wells & Xenias, 2015).

Within this overall pattern, however, there is considerable heterogeneity. In Switzerland, for example, French-speaking municipalities have 3–6% more energy-efficient vehicles than German-speaking ones, despite sharing the same sets of institutions. Research suggests that this may relate to French speakers having a stronger sense of collectivism and altruism (Filippini & Wekhof, 2021). People also have different driving cultures: Japanese drivers are more energy efficient because they are more strongly focused on safe driving and risk reduction, while US drivers prioritise individual freedom, which is associated with greater risk-taking while driving and more frequent crashes (Sovacool & Griffiths, 2020). Across European countries, considerably different mobility cultures have been identified in sub-regional clusters, which suggest very different policy interventions would be needed to achieve sustainable mobility (Haustein & Nielsen, 2016).

These studies all compared national or linguistically defined cultures, but cultural differentiation is equally able to be explored through other dissections. For example, although the evidence for this is mixed, recent studies suggest that some younger people in industrialised countries are less likely to want to own a car than their peers or previous generations (Delbosc, 2017). This trend was explored qualitatively in a study of 51 young people in New Zealand. Around half had a driver's licence and owned a car, while the rest had made a choice not to. For the car-owning group, driving was the default choice for mobility, and most disdained other transport modes. They spoke of their cars as giving them freedom and independence, safety, job opportunities and a social life. In contrast, those who had chosen not to get a licence or own a car generally used multiple forms of mobility: they walked, biked, took buses and used shared transport. Some gave an environmental rationale for their choices, but for others it was about saving money or was simply the easiest choice. Surprisingly, they frequently referred to the *absence* of a car as giving them freedom and independence (Hopkins, 2017; Hopkins & Stephenson, 2016).

The New Zealand study also elaborated on the wider influences that were reinforcing car ownership among young people such as roading infrastructure and urban design, investment policies, fuel subsidies and the social status afforded to car ownership, as illustrated in Fig. 5.1. Recently emerging influences that were tending to support a multi-mobility culture included new policies and investments supporting walking and cycling, and growing health and environmental concerns across society (Hopkins & Stephenson, 2016).

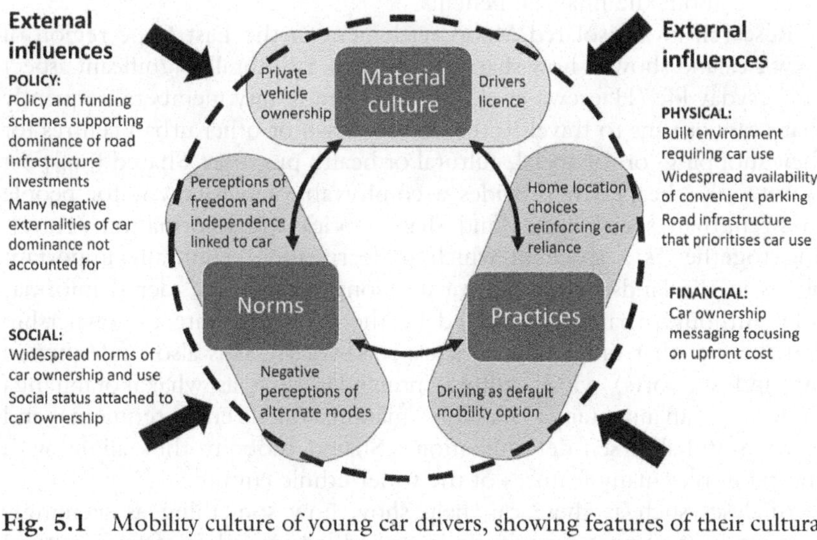

Fig. 5.1 Mobility culture of young car drivers, showing features of their cultural ensemble (within the dashed circle) and external influences (beyond the dashed circle) (Reproduction of Fig. 2 from Hopkins and Stephenson [2016])

This divergence in mobility cultures among young people offers some insights into the process of transition towards a low-carbon mobility system, which by necessity will require far greater use of active, public and shared transport. The memberships of multi-mobility culture groups will need to grow rapidly, so it will be important to understand why some actors are not conforming to the dominant culture of automobility, and how this cultural divergence can be maintained.

One of the solutions often put forward to reduce the proliferation of cars is shared mobility, which involves sharing the use of a single vehicle between multiple people. This can be effected through means such as co-ownership, car-share and bike-share programmes, ride-hailing and other mobility-as-a-service offerings. From a cultural perspective, shared mobility challenges established norms, especially those that prioritise individuality and the symbolism of personal ownership (Axsen & Sovacool, 2019). Yet shared transport has a long history and is still undertaken in many less-wealthy communities in ways that do not require technologically 'smart' systems. Looking at other systems of mobility can help reveal how 'cultural' all mobility choices are, and that ridesharing does not have to be all about the financial benefits.

Research in an isolated Māori settlement in the East Cape region of New Zealand showed how shared mobility is a culturally significant aspect of everyday life (Haerewa et al., 2018). Community members frequently share cars or vans to travel to the nearest town or other urban centres for their shopping, or for social, cultural or health purposes. Shared transport is not only cheaper, it provides a comfortable environment for people to strengthen social bonds and share stories and information. Travelling together is a space in which to learn about culturally important places in the landscape, learn waiata (songs) and share sacred information. Although sharing is spurred by the cost of private car ownership and the fact that only some cars are roadworthy, it is also underpinned by (and supports) wider cultural principles such as whakakotahitanga (unity), whanaungatanga (family), māramatanga (enlightenment) and mana motuhake (self-determination). Shared mobility thus aligns with and reinforces many features of the wider ethnic culture.

Analyses such as these can help show how something as seemingly innocuous as using a private car is embedded in and reinforces cultural processes. Any mobility system–whether it is dominated by private car use, multi-mobility or shared transport—is culturally embedded as well as having distinct social, environmental and economic outcomes.

Identifying Cultural Patterns

Depending on the sustainability outcome of interest, it is usually possible to identify clusters of actors within a population that have similar cultural ensembles and thereby similar outcomes. Here are four examples of studies that identified such clusters.

Energy Cultures in Transylvania

For some sustainability questions, there will be an easily identifiable dominant culture: a widely shared and largely similar ensemble of motivators, activities and materiality. In a rural area of eastern Transylvania in Romania, researchers found that most households shared a similar energy culture. Households typically had few appliances and frugal everyday practices. There was a widespread reliance on wood for heating and cooking, and around half the homes supplemented this with propane gas. Families saw themselves as stewards of the local environment and generally sought to conserve energy and other resources. The ubiquity of this cultural ensemble was statistically identified from survey data as well as through observations by the research team (Klaniecki et al., 2020).

The researchers were exploring the potential for new low-carbon energy systems in the region. They found that the Transylvanian households were strongly interested in switching to renewables, especially solar. Based on the cultural analysis, they concluded that any new energy system should be designed to fit the prevailing culture: a reliable and affordable energy supply that supported the shared sense of stewardship and aligned with households' norms including their interest in local resource use.

Energy Use in Older Households in Italy

In some populations, there may be more than one distinct cultural ensemble that relates to a given outcome. Sometimes cultural differences may arise from intergenerational divergence. In Italy, for example, researchers used a cultural analysis to explore why the now-elderly war generation used far less energy than the baby boomer generation. Using data from 22,000 households, they looked at changing energy expenditure over time as well as dwelling features and household characteristics. They found that the difference in expenditure on energy was less an age effect, and more to do with a cohort effect that was distinctly cultural. War generation Italians were more likely to be frugal in their old age while baby boomers were more profligate, had more energy-consuming appliances, greater use of heating and air conditioning, and less focus on energy saving. The researchers concluded that this intergenerational shift in energy cultures would have significant implications as the more energy-intensive cohort aged, as consumption tends to peak between 50 and 60 years. They concluded that if the effects of age and intergenerational difference in energy culture was not accounted for, it would result in a severe under-estimation of future national energy demand (Bardazzi & Pazienza, 2017, 2020).

Energy Efficiency Among New Zealand Households

Another study found divergent culture groups that had different energy efficiency outcomes (Lawson & Williams, 2012). Using data from a demographically representative sample of 2300 New Zealand households, the study identified four distinct clusters. The 'Energy Efficient' cluster had efficient practices as well as efficient houses and appliances, while 'Energy Easy' had relatively efficient material items but practices that were not particularly efficient. The 'Energy Economic' cluster of households tended to have inefficient material items yet very efficient practices, using relatively little energy—these were generally lower-income households. 'Energy Extravagant' tended to be higher-income households with inefficient practices, and with many energy-hungry or relatively inefficient appliances. This last group used the most energy. Many 'Energy Economic' households had young adults, and many in the 'Energy Efficient' group were couples whose children had left home, and the authors concluded that the clusters were related to some extent to life stage and to household income, but that these were not complete explanations. Based on the clustering, the research team recommended policy interventions for improving energy efficiency that were specific to each cluster (Barton et al., 2013).

These last two examples used statistical methods to identify distinct cultural clusters across a population, using a relatively small number of variables. Other studies have used qualitative methods to identify cultural differences within a population under study. For example, gender-related differences in energy culture identified through qualitative research include the habituated roles of males and females in relation to energy use (Jürisoo et al., 2019); gendered roles in energy-related decision-making (Bach et al., 2020); and the different effects of energy transitions on men and women within a given societal group (Johnson et al., 2019).

Cultural Resistance to Change

Established cultures can be remarkably stable even if there are considerable adjustments in the context within which actors operate. From a sustainability perspective, this is a beneficial quality to have if a culture is already sustainable, but cultural stability can be a problem if it means that interventions that are intended to improve social or environmental outcomes are not successful. The examples in this section reveal various cultural dynamics that have a role in resistance to change.

Household Energy Efficiency in Norway

Energy efficiency is arguably a rational choice because it saves people money, but cost is only one of many influences on energy decisions. Aspects of people's existing cultural ensembles, such as beliefs, norms, knowledge and routines, can operate to maintain the status quo and trump the influence of price signals. Evidence of how cultural characteristics resist change can be seen in research that explored Norwegian households' rationalisations and norms relating to energy efficiency (Godbolt, 2015).

Efficiency can be achieved in two main ways—through more efficient practices, or through the adoption of energy-efficient technologies. Drawing from focus groups, the researchers found that the Norwegian households largely failed to do either. Most households were not efficient in their energy use, and even those who were most price-conscious still used as much energy as they needed to remain comfortable. When it came to energy-efficient practices, householders mainly chose to undertake activities that easily integrated with their daily routines, and they avoided actions that involved hard work or were time-consuming. With regard to purchasing energy-efficient technologies, many participants did not feel that the cost savings were sufficient to motivate investments, and often used upfront cost as a reason not to invest. Where households had invested in efficient technologies, they usually explained it using non-cost rationales (e.g. better and more stable heating).

To investigate why this was the case, the researchers focused on the rationalisations and norms that were part of this inefficient energy culture. They found that the householders' actions were underpinned by four principles of conduct that were repeatedly mentioned when people were asked to explain their energy use. These moral positions were often in tension or conflict. A morality of saving (i.e. valuing thriftiness) was moderated by a morality of merit (i.e. saving energy in one aspect of everyday life allowed profligate energy use elsewhere), a morality of needs (i.e. all of household energy consumption is needed to carry out everyday activities) and a morality of entitlement (i.e. a right to use as much energy as needed without justification). The ultimate outcome of this internally conflicted 'ethos' was that households were not particularly energy efficient even though it would have saved them money (Godbolt, 2015). This work reinforces the importance of understanding the motivators that shape people's routines and investment choices, as these ultimately shape sustainability outcomes. Using the terminology of the revised cultures framework, it illustrates how disjunctions between motivators can prevent change in other aspects of culture (see Fig. 4.2 in Chapter 4).

Cooking in Zambian Households

In Zambia, interventions by the government and by non-governmental organisations were failing to persuade households to replace their charcoal stoves with clean, low-cost cooking methods such as electric stoves or fuel-efficient biomass pellet cookstoves (Jürisoo et al., 2019). This attempt to alter household cooking methods was in part because the charcoal stoves cause respiratory problems, especially among women and children, who spend more time indoors. Additionally, demand for charcoal was a cause of widespread deforestation, and charcoal prices were high because of the consequential supply issues. Sector experts considered that because alternative cooking methods were cheaper and had no supply issues, households would be keen to switch. However, families largely resisted change and the take-up of other cooking methods was low.

Using the cultures framework to analyse why this was the case, the study found a strongly habituated cooking culture whereby the charcoal stove was central to Zambian identity. 'This is how cooking is done in Zambia: it connects us to our roots', explained one woman (Jürisoo et al., 2019: 62). Use of the charcoal stove gave food the desired traditional flavour. Because it didn't require ongoing supervision, the stove also allowed women to carry out other household chores alongside cooking. Going out to get charcoal was an opportunity for socialisation and to catch up on the latest news. The charcoal stove was thus integral to patterns of behaviour, socialisation and expectations around food. Despite the high price of charcoal, the apparent advantages of other cooking technologies weren't enough to trigger change.

Relating this to the cultures framework, the researchers concluded that households had very little agency, and were largely locked into cultural ensembles involving norms around food and daily routines, the material culture of cookstoves and food, and practices including cooking, housework and socialisation. These were mutually reinforcing and further reinforced by external factors such as the presence of charcoal sellers in nearby streets, electricity shortages and national policies which included a lack of monitoring of illegal charcoal supplies. Together, the strongly linked cultural ensemble, lack of agency and reinforcing external influences meant that the cooking culture was highly resistant to change.

Energy Efficiency in the US Navy

Similar resistance can be seen in an example from the US Navy (Dew et al., 2017). Efforts had been made to improve lighting efficiency in the fleet for at least 12 years through navy programmes, but nearly all the ships still used inefficient lighting despite the technical, safety and cost advantages of LED lights. Many different justifications (e.g. less labour, reducing fuel consumption, improving fighting capability, reducing overall energy expenses) had been used to support the change, but these had gained little traction.

Using the cultures framework, the research revealed aspects of the navy's norms, materiality and practices that were strongly aligned, and interacted to support the existing inefficient and costly lighting system. This included beliefs about energy being cheap and abundant and the navy's mission being war not efficiency; material aspects like existing fixtures and their maintenance requirements; and various practices such as acquisition processes and officer rotation. These are summarised in Fig. 5.2. The external influences on this culture are elaborated later in this chapter.

Together, this system of interlocked cultural influences was preventing the navy from making a very simple change to LED lights that would be

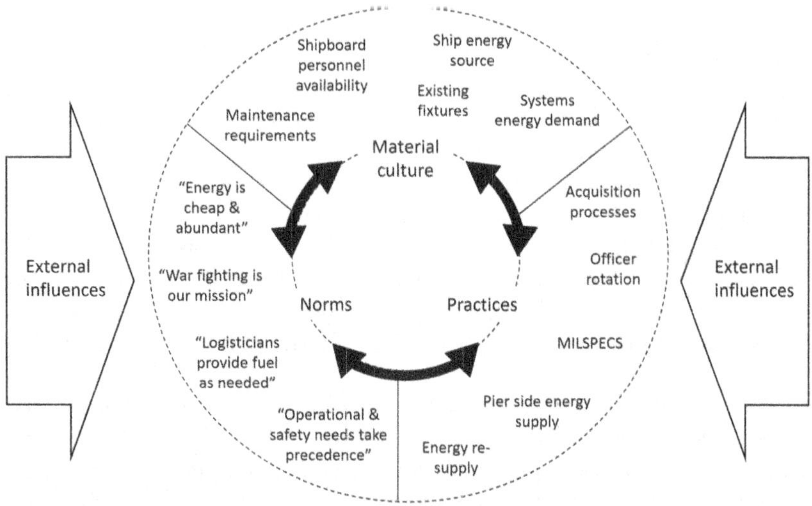

Fig. 5.2 Internal elements of US Navy energy culture (Reproduction of Fig. 2 from Dew et al. [2017])

cheaper because they would be longer lasting and require far less energy. The study authors looked at eight alternative explanations for the sluggish adoption of more efficient lighting technologies and concluded that the most plausible explanation was 'the difficulty of crafting a good justification for adoption, one that has a favourable fit with the prevailing energy culture' (Dew et al., 2017: 64).

Housing Retrofits in Ireland

In some instances, cultures can be impervious to change even where there is a significant alteration in context. This resistance is illustrated by a study in Ireland which showed why the intended benefits of an energy efficiency makeover in a social housing estate were not fully achieved. Prior to changes to their houses, householders experienced indoor condensation and mould growth, and were concerned about how cold and draughty their homes were. The intervention included installing wall and ceiling insulation, double glazing, ventilation systems and new heating systems with thermostats that could be manually over-ridden. Using the cultures framework, the study followed 20 households before and after the retrofits, looking at changes in material conditions, energy-related practices and in householders' attitudes, perceptions and norms (Rau et al., 2020).

The research found that the physical retrofit resulted in some improvements in thermal comfort, but not to the extent intended by the designers. They concluded that this was because of the persistence of inefficient routines around heating and appliance use, and also because the householders lacked the knowledge and practices required to successfully operate the new water and space heating systems. Despite the intention that heating would be automated, many of the households continued to operate the heating system manually as they had done in the past. They set the thermostat high and physically turned it on and off rather than using the more efficient option of setting the thermostat at a lower level and letting the adjustments occur automatically. Only half of the households increased the average temperature of their houses despite having previously been unsatisfied with levels of warmth.

The cultures framework provided a helpful basis for integrating physical measures and qualitative information to illustrate variations between households in whether and how their energy culture changed following the retrofit, and how this related to changes in gas and electricity consumption. The average outcome was a 19% reduction in total energy use—a positive result given many households were economically vulnerable and were able to increase the comfort of their homes. However, this varied between households: the energy savings were lower than expected, and the energy rebound absorbed any further potential financial savings. The material changes to hot water and space heating resulted in modest changes in related practices but did not shift the prevailing energy culture in any fundamental way, and nor did it result in more efficient use of other appliances.

The authors concluded that efficiency retrofit programmes should put equal effort into analysing and, if necessary, seeking to adjust all aspects of a household's energy culture, not just the material aspects. It is equally important to pay attention to householders' energy-related expectations, aspirations, understandings and practices if real and lasting reductions in energy use are to be realised along with health and wellbeing benefits (Rau et al., 2020).

Cultural Dynamics in Switzerland

Household adoption of solar photovoltaics (PV) is arguably one route to lower-carbon electricity, and many nations have used subsidies to encourage their uptake. However, households vary in their willingness to adopt PV. A study of households in Berne, Switzerland, looked at how cultural attributes influence the uptake of PV (Bach et al., 2020). Interviews with adopters and non-adopters showed that they were quite alike in many aspects of their 'electricity cultures': both groups engaged in similar energy behaviours in the home, both perceived electricity efficiency as important and both perceived PV as a reliable technology. However, adopters of PV showed some cultural differences—they were more environmentally motivated, aspired to energy independence, did more to reduce electricity use and generally owned more renewable technologies.

Non-adopters' rationalisations of their positions revealed how their cultural dynamics essentially closed out the potential for PV adoption through the strong reinforcing linkages between their norms, practices and materiality, while adopters had somewhat different cultural ensembles and more flexible links between cultural elements. This is illustrated in Fig. 5.3, which uses quotes from the non-adopter interviews to indicate how the cultural dynamics create a self-sustaining system that hinders PV adoption. The authors concluded that this cultural analysis shows the need for a shift from homogenous financially focused policies encouraging PV uptake, to policies that account for cultural difference across households (Bach et al., 2020).

Cultural dynamics can thus complicate or inhibit interventions designed to improve sustainability outcomes. This is further demonstrated by a review of interventions for more sustainable forms of mobility and energy use in 28 countries (Sovacool & Griffiths, 2020). Mobility examples included aggressive driving, speeding and eco-driving, automated vehicles, and ridesharing and carpooling. Energy-related examples included solar home systems, improved cookstoves, and energy-efficient heating, cooling and hot water practices. The findings revealed how cultural ensembles can play as significant a role as price signals, national programmes and regulations in the adoption of new technologies and more efficient behaviours.

The review paper showed that impeding cultural factors included social customs, conspicuous consumption to project wealth or power, peer pressure, spiritual beliefs, traditional practices, gender roles, and

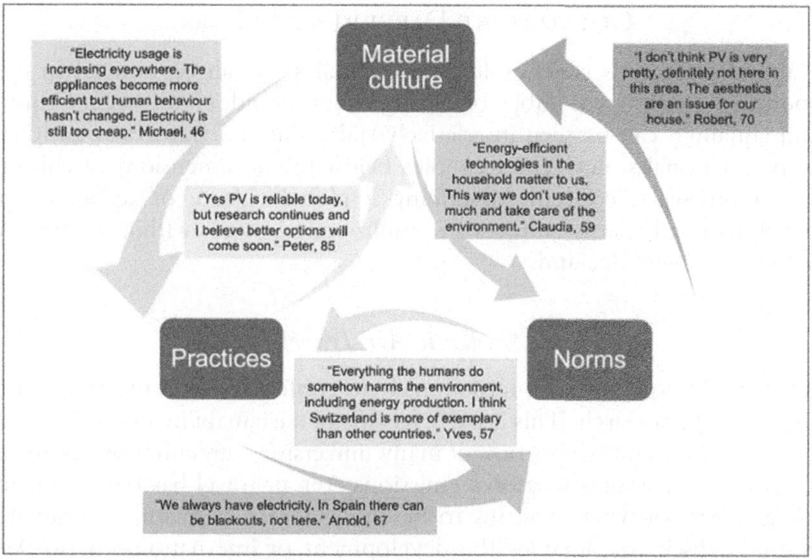

Fig. 5.3 Dynamics of a non-adopter culture. The positive interplay between cultural attributes creates an ensemble that is resistant to change, hindering PV adoption (Reproduction of Fig. 3 from Bach et al. [2020])

misperceptions and biases in technology design. The authors concluded that '[s]ome emerging energy and low-carbon innovations can create, challenge, or reinforce existing cultures; in other situations embedded cultures can challenge, shape, and entrench particular low-carbon innovations and practices. Ideas, customs, and social behavior merge with technological artefacts and material infrastructures to create cultures (or sub-cultures) of driving, automation, riding, domesticating, cooking, and heating' (Sovacool & Griffiths, 2020: 7). Culture is often overlooked in policy development for low-carbon transitions which are usually designed on techno-economic considerations and assumptions about individual decision-making processes. The authors concluded that policy would be much more effective if it accounted for the ways in which cultures shape aspirations, capabilities and agency for low-carbon transitions.

Culture at Different Scales

Cultural stability is observable at many scales. It can be noticed in the enduring cultural ensemble of a single household or an organisation, and equally it can be seen in relatively stable cultural features at national or pan-national scales. The examples below reveal dimensions of culture as a multi-scalar concept, describing a global culture of academic air travel, national energy cultures and multi-scale cultures within the timber industry in New Zealand.

Academic Air Travel

Academics frequently fly internationally to conferences and meetings, and to undertake research. This activity has major sustainability downsides—it causes a significant proportion of many universities' greenhouse gas emissions and contributes to global emissions. Yet air travel has long been an integral part of what it means to be an academic. It maintains visibility within a discipline, enables the development of international networks, and supports international collaborations and jointly authored papers. These are regarded as key measures of academic success and can affect promotions. Furthermore, comparative rankings of universities take into account the international outlook of individual universities, which is in part assessed through the proportion of a university's academic publications with at least one international co-author. International travel, including flying, is therefore an expected part of scholarly life.

Tseng et al. (2022) described this as an academic transport culture and were particularly interested in how the COVID-19 pandemic destabilised this culture (possibly temporarily). Academic transport culture involves scholarly norms about career development, network building, sharing of research, keeping up with academic advances, research collaborations and co-authored papers. These norms align with the practice of flying frequently, and this has been further encouraged by funding available through universities to support international travel. Transport culture can be seen to be operating in at least two scales of actors—at the scale of academics and at the scale of universities.

With COVID-19, academics were forced to adopt new practices (virtual conferences, online meetings) and new materialities (technologies and software that allow stable high-speed videoconferencing) which have resulted in the normalisation of virtual meetings and conferences and the new forms of social interaction that they involve. Generally, positive experiences with these virtual interactions have shown that the need for academic flights to support scholarly life may not be as great as assumed. The authors suggest that in a post-COVID-19 environment, these temporary culture changes could be made more permanent to maintain the reductions in greenhouse gas emissions. This should include changes in transport culture at the university scale, including adjustments to institutional expectations for academics to travel, institutional investments in quality videoconferencing equipment and new practices around the organisation of conferences.

National Energy Cultures

As already touched on above, sustainability-related cultures among citizens can differ greatly between nations. Cultural differences can also be identified at a state level. One example is a study that sought to explain the significant variability in countries' decarbonisation ambitions relating to the Paris Agreement (Stephenson et al., 2021). Prior research on this question had failed to establish any single compelling reason for the wide variance in countries' nationally determined contributions and their intended low-carbon pathways. Previous comparative studies had generally each focused on only one or two potential explanatory factors for this divergence. These included countries' natural resource endowments, political structures, levels of wealth or development, current dependence on fossil fuels, political leadership, job creation or geopolitical ambitions. This study used the cultures framework as an organising structure around which to explore these and other factors in an integrated fashion.

Using case studies of India, Denmark, China and Russia, the authors investigated cultural (and other) influences on energy policy evident in each nation over a 30-year period. They concluded that national low-carbon ambitions are contingent upon how a nation perceives the role of energy, and the choices, policies, investments and actions that flow from this. These interactions create a cultural 'system' that comprises the interplay between normative, material, institutional and policy-related attributes of national decision-making bodies.

The habituative nature of this system shapes the extent to which nations are willing to respond to the global challenge of climate change. For example, Denmark's energy culture was described as being proactive and innovative for low-carbon wellbeing, while Russia was characterised as having a hegemonic energy culture, focused on geopolitical dominance and economic restoration. By revealing the heterogeneity of national energy cultures, the study showed why their responses to the climate crisis are so different. It concluded that because of this cultural variability, each nation would require different stimuli to strengthen its low-carbon ambitions (Stephenson et al., 2021).

Multi-Scale Culture in the New Zealand Timber Industry

A study in New Zealand looked at the barriers to the use of energy-efficient drying technologies in the timber industry. Larger firms preferred vented kiln dryers (less efficient and producing significant particulate emissions) over the newer heat pump kiln dryers (more efficient and no emissions) even though there were no significant differences in the average operating costs, drying costs or commercial success. The newer technology offered the advantages of energy-efficient drying and better-quality wood products, and would help firms avoid the risks of path dependency in an increasingly emissions-conscious world. The study identified culturally based resistance to change among individual firms and across the timber drying sector as a whole (Fig. 5.4) (Bell et al., 2014).

At the sector level, the dominance of vented kilns was strong supported by the prevailing technologies, practices and norms, including research investments and the focus on this technology at industry-wide events. The sector-wide culture hindered the adoption of innovations and technological learning at the scale of individual firms. The sector culture acted as an 'external influence' on the energy cultures of individual firms, constraining their choices and normalising the traditional kiln systems. Influential stakeholders in the industry were embedded in the dominant energy culture, while the few smaller firms who had adopted heat pump dryers had little influence in the wider sector. The researchers concluded that this multi-level culture created such a strongly self-replicating system that any transformative change would need to be initiated from outside the sector (Bell et al., 2014).

These examples show how the cultures framework can be used as an analytical framing regardless of the actor or the scope of their influence. The academic travel and timber drying examples reveal how cultures can operate at different scales, each with its particular ensemble of motivators, activities and materiality, with the broader-scale culture shaping the cultures of less powerful actors. The study of national energy cultures

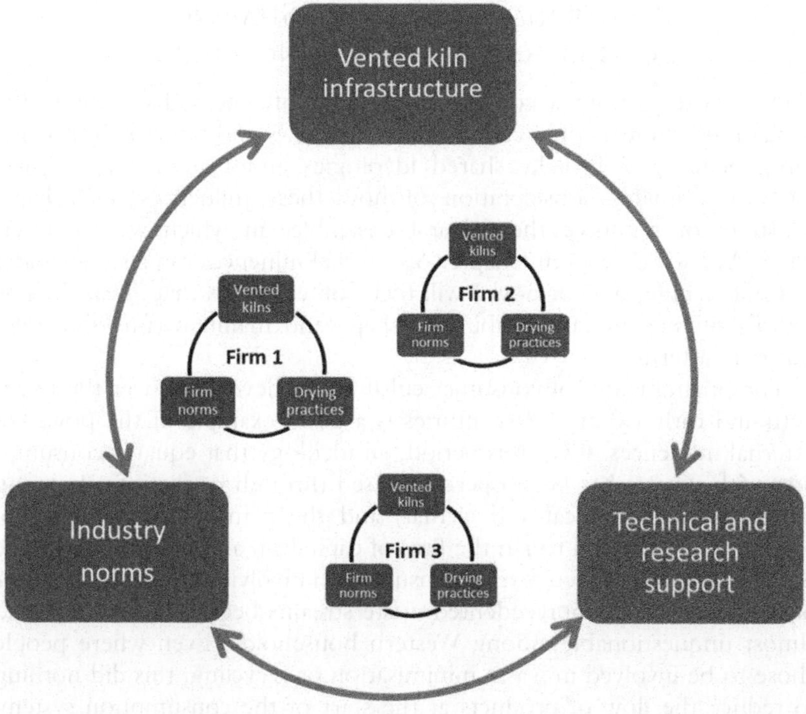

Fig. 5.4 The energy culture of individual firms is shaped and constrained by the prevailing energy culture of the timber drying sector (Reproduction of Fig. 4 from Bell et al. [2014])

shows how cultural qualities can evolve and consolidate over time, and that cultural expressions regarding a resource like energy can be highly diverse. Culture's tendency to durability is notable in all the examples, apart from way in which COVID-19 disrupted academic transport culture. There, a significant change in context reset the culture into new patterns, some of which may continue to endure even now that international travel restrictions are lifted.

External Influences Shaping Culture and Constraining Change

Cultures exist within a contextual soup of influences that include the geographic context, political arrangements, laws and policies, infrastructure, media, and broadly shared ideologies and beliefs. The cultures framework invites consideration of how these influences may shape, constrain or reinforce the cultural ensembles in which we are interested. As I will discuss in Chapter 6, externalinfluences can also engender cultural change, but for now I will focus on examples that illustrate how external influences can combine to shape and maintain (unsustainable) cultural patterns.

The phenomenon of consumer culture that developed over the twentieth and early twenty-first centuries is a good example of the power of external influences. Over this period, an ideology that equates consumption with success has been operationalised through marketing, shopping infrastructures (physical and virtual) and the planned obsolescence of products, all designed to fan the fires of capitalism and increase corporate profits. The resulting culture of consumption involving the rapid turnover of possessions and unprecedented waste streams became normalised and almost unquestionable among Western households. Even where people chose to be involved in waste minimisation or recycling, this did nothing to reduce the flow of products at the start of the consumption system. Households became a conduit of matter that was being transformed from resource to consumer item to trash. The outcomes are well understood, from plastic gyres in the ocean, e-waste dumps in developing countries and microplastic pollution in soils and water. Even with a greater awareness of the environmental and social costs, most households are still largely locked in to these problematic patterns of consumption by structures and ideologies that are beyond their control (Davies et al., 2014).

The cultures framework helps to identify and depict how external influences, including broader-scale cultures, shape the culture of the actors in whom we are interested. Most of the studies I have discussed so far in this chapter have identified external influences that have shaped the relevant culture/s or are constraining change. In the US Navy example, a range of external influences maintained its inefficient energy culture. These included Congressional appropriation processes, the energy culture of the Department of Defense (which differed from that of the navy), executive actions of the Federal Government, a limited supplier base, formalised instructions and the cost of energy relative to other costs (see Fig. 5.5). In order to undertake a simple action of investing in LED lightbulbs, naval personnel were faced with justifying this action through complex processes to other more influential organisations with little interest in acting outside of business-as-usual.

External influences on culture often involve power differentials. Influential organisations and individuals, systems of governance, infrastructure, ideologies, media and resource allocations can all shape less powerful cultures. Because of this, we should never assume that people have necessarily freely chosen to adopt cultural attributes. These may have been absorbed and learned over time from family and peers, but equally they may also be imposed and reinforced by structural forces beyond people's control. This is illustrated in the following examples.

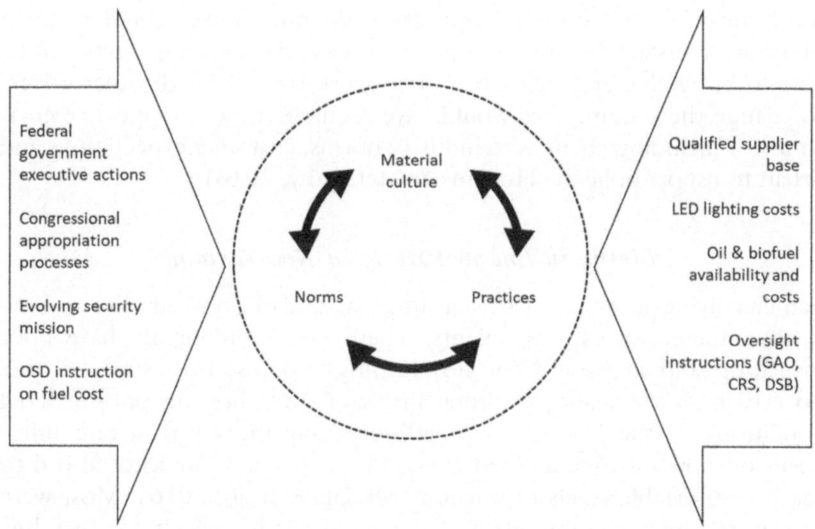

Fig. 5.5 External influences on US Navy energy culture (Reproduction of Fig. 1 from Dew et al. [2017])

Consumer Expectations of Urban Freight Deliveries

A study in the urban freight sector in New Zealand revealed how changing external influences were causing freight deliveries to become less efficient despite many freight businesses wishing to become more efficient and sustainable. The sector was being fundamentally changed by the impacts of escalating online orders, leading to increasing home deliveries, new tracking technologies and rising customer expectations of tracking and delivery times. Together with freight firms de-risking through subcontracting deliveries to owner-drivers, this meant freight drivers had to be highly competitive and make multiple journeys with low-bulk deliveries. There were significant social, economic and environmental outcomes: owner-drivers faced high stress, low profitability and lack of security, and the delivery system generated high fuel use and greenhouse gas emissions and contributed to increased congestion on urban streets. The owner-drivers interviewed felt they had little option but to meet consumer expectations, and although some drivers could envisage other more sustainable ways to operate, they had insufficient agency to be able to make changes themselves or to work together with other drivers to change the system. This would have required a more supportive environment including changes to industry norms, customer expectations and urban transport policies (Hopkins & McCarthy, 2016).

Living in Energy Poverty in New Zealand

Families living in energy hardship in New Zealand typically live in low-quality houses, usually as tenants. These homes frequently have poor insulation and inefficient (or no) heating systems. In a study of fuel poverty in New Zealand, a cultural analysis showed how the poor material conditions of the houses and inefficient appliances had strong influences on how inhabitants used energy, because they could not afford to reach comfortable levels of warmth (McKague et al., 2016). Most were in private rental accommodation, the standards for which are very low compared to most European and North American countries. Even where people owned their homes, they lacked agency to fix them. For a variety of reasons the households were on low incomes or had additional costs such as illness or large families. There was an absence of state support to improve housing standards apart from an insulation subsidy that was not available to landlords. An electricity pre-payment system that was intended for households on low incomes turned out to be more expensive than standard payment systems.

Given these circumstances, common practices included avoiding using heating appliances and staying in bed for extended periods to keep warm, or heating only one room of the house, thus confining activities to that room. Family members often suffered ill health, and also were socially isolated because they were ashamed to bring others to their cold house. Some spent much time scavenging for free firewood to keep their stove burning. They experienced high levels of anxiety about covering their energy costs, and this competed with their ability to afford food. This energy culture was the only response possible by the households due to circumstances that were almost entirely beyond their control. For these families, there was no way out of their frugal and unhealthy energy culture, so these characteristics will likely endure unless there is a change to the families' agency and/or the material conditions in which they live.

Distributive Injustice in Slum Rehabilitation Housing in Brazil, India and Nigeria

My final example is an investigation of distributive injustices in slum rehabilitation housing in the cities of Mumbai, Abuja and Rio de Janiero (Debnath et al., 2021). In each city, the families had been rehoused into a distinctive type of rehabilitation housing and provided with appliances. The researchers undertook focus group discussions with women in each location and designed the questions around the elements of the cultures framework. The study found that their energy practices, norms and materiality related strongly to the type of building and energy appliances they had been allocated. In Abuja, for example, the housing units were single storey with common spaces, and these allowed for community ownership and shared use of appliances. The high-rise buildings in Mumbai meant practices such as cooking and cooling which were energy intensive due to a lack of outdoor and shared spaces. In Rio de Janiero, energy practices were shaped by appliances that had been donated by higher-income groups and which were not necessarily a good fit with cooking traditions.

Other external influences compounded the energy injustices experienced by these households. For example, in Ajuba the families faced high and irregular power bills, low-power quality and frequent load shedding that damaged their appliances, which were hard to repair or replace. They also suffered health impacts as they often used firewood when the power was down, leading to indoor pollution from smoke. In Rio, the donation-based model passed on inefficient appliances with the associated burden of higher operational and maintenance costs, and poor housing design meant that significant energy needed to be spent on cooling. The researchers concluded that the families in each case suffered from structurally derived injustices, where the external influences on households maintained them in a situation of energy poverty. The imposed energy cultures were almost impossible for the household members to alter.

Culture's Tendency to Resist Change

As shown by these examples, some cultures can exist in a state of relative stability with very little change over long periods of time. Sometimes this can be positive for sustainability ambitions (e.g. the retention of Indigenous knowledge and practices through shared mobility), but most of the examples focus on negative implications, such as for health and equity (e.g. households locked into energy poverty), energy efficiency (e.g. timber firms, Norwegian households) or greenhouse gas emissions (e.g. car culture, academic transport culture).

The dynamics between elements in cultural ensembles play a role in this continuity. With the Zambian example, charcoal cookstoves were strongly tied to the food flavours that people enjoyed and set a pace of cooking that enabled other tasks to be carried out simultaneously. The purchase of charcoal from street sellers was an important part of social life. The comfortable normalities of these routines and foods, and their central role in creating a sense of community, meant that people strongly resisted a change in cooking technologies. The lock-in was a result of the strong links between motivators, activities and the material items of the cookstoves and charcoal. Similar dynamics between elements of cultural ensembles were seen in the Switzerland example of non-adopters of PV.

The Norwegian example shows a different situation, where resistance to change to more efficient household practices arose from internally conflicting motivators for their energy use. Despite valuing thriftiness, they simultaneously held other justifications for using as much energy as they needed or desired. The result was that the householders' inefficient energy cultures were maintained.

Culture can be so tenacious that cultural traits continue even where some cultural elements change. This is illustrated in examples where alterations to the material aspects of people's lives failed to alter everyday activities to the extent anticipated. In Ireland, some households continued with their old routines despite new heating systems, thus failing to benefit fully from the promised financial and efficiency gains. In Italy, the war generation cohort was still far more frugal with energy use than the baby boomer generation, carrying norms and practices from the past throughout their lives despite positive changes in wealth and housing quality.

The examples also reveal the role of external influences in constraining cultural change (as with the US Navy example), forcing cultures to become more unsustainable (as with the urban freight example) and impelling actors to adopt cultural ensembles that are not only unsustainable but also threaten the actors' health and wellbeing (as with the fuel poverty and slum rehabilitation examples). In each case, the actors had limited agency and, due to power differentials, would have little chance of altering their own cultural ensemble, let alone altering the external influences that were shaping their particular energy culture.

External influences can be cultural, and the examples also show how more dominant and widespread cultural ensembles can act as an external influence on other less dominant cultures. Such multi-scale cultures can be identified where certain actors' cultural attributes influence the cultural attributes of others with less agency, such as the energy culture of the timber sector compared to that of individual firms, or the energy culture of landlords in relation to tenants. In these instances, the more powerful culture acts to shape (and often constrain) the cultures of the less powerful.

Cultural attributes are rarely adopted by actors in a conscious and purposeful way. Cultural ensembles—ways of thinking, acting and having—are learned from others over time, often relatively unconsciously. They are also a response to the conditions in which people find themselves. Where cultural actors lack power, they can become accustomed to patterns of behaviour and material choices, and this too can constrain them from changing their culture despite their aspirations to change, as with the examples of freight drivers and households in energy poverty.

Conclusion

When one is inside a culture, it is hard to see it as anything but 'just how life is'. It can be even harder to identify the dispersed influences that shape one's culture. For those who have always been dependent on cars, have always lived in cold houses or have always operated within the US Navy hierarchy, their particular cultural ensembles may appear normal and largely unquestionable. It seems to make little difference to the longevity of a culture whether a cultural actor is unhappy with the ensemble of which they are part (e.g. households in energy poverty) or entirely comfortable with it (e.g. most car users). Making culture visible, including to cultural actors themselves, is one of the challenges of achieving a sustainable future.

The examples in this chapter have explored cultural characteristics in relation to a wide variety of sustainability-related outcomes, including energy efficiency, health and wellbeing, energy consumption, energy poverty, transport emissions, retention of cultural values, uptake of new technologies and distributive injustice. The examples underline that what culture is for one question is different to what it is for another question. A household's cultural ensemble relating to water consumption will likely be very different to their cultural ensemble relating to mobility: both the cultural ensemble and the membership of the respective culture groups are likely to differ. The energy culture of a business will comprise a very different set of features to the energy culture of a nation state. Yet all of these cultures can be described using the language and structure of the cultures framework. The framework thus offers a universal, scale-free model for the analysis of culture.

Applying the cultures framework to investigate relatively stable cultures has revealed several factors involved in cultural stasis. These include situations where cultural elements are closely and positively linked (as with the Zambian cookstove example), situations where actors hold internally inconsistent motivations or rationalisations (as with the Norwegian household example), situations where actors have limited agency to change even if they wish to (as with the energy poverty and urban freight examples) and situations where external influences, including more powerful cultures, shape and continually reinforce cultural ensembles (as with automobility). These dynamics can operate singly or collectively to reinforce cultural stability.

In this chapter, I have explored how culture can become habituated and relatively difficult to change. Cultural habituation is not necessarily a bad thing in sustainability terms. Some cultural patterns have positive sustainability outcomes and should equally be studied to understand their longevity. But cultural habituation is a problem when humans are caught up in unsustainable patterns of consumption and production. In the following chapter, I use research-based examples to discuss a more positive topic—how and why cultural change occurs.

References

Axsen, J., & Sovacool, B. K. (2019). The roles of users in electric, shared and automated mobility transitions. *Transportation Research Part D: Transport and Environment, 71*(March), 1–21.

Bach, L., Hopkins, D., & Stephenson, J. (2020). Solar electricity cultures: Household adoption dynamics and energy policy in Switzerland. *Energy Research and Social Science, 63*(October 2019), 101395.

Bardazzi, R., & Pazienza, M. G. (2017). Switch off the light, please! Energy use, aging population and consumption habits. *Energy Economics, 65*, 161–171.

Bardazzi, R., & Pazienza, M. G. (2020). When I was your age: Generational effects on long-run residential energy consumption in Italy. *Energy Research & Social Science, 70*, 101611.

Barton, B., Blackwell, S., Carrington, G., Ford, R., Lawson, R., Stephenson, J., Thorsnes, P., & Williams, J. (2013). *Energy cultures: Implication for policymakers*. Centre for Sustainability, University of Otago.

Bell, M., Carrington, G., Lawson, R., & Stephenson, J. (2014). Socio-technical barriers to the use of energy-efficient timber drying technology in New Zealand. *Energy Policy, 67*, 747–755.

Davies, A., Fahy, F., & Rau, H. (2014). *Challenging consumption: Pathways to a more sustainable future*. Taylor & Francis.

Debnath, R., Bardhan, R., Darby, S., Mohaddes, K., Sunikka-Blank, M., Coelho, A. C. V., & Isa, A. (2021). Words against injustices: A deep narrative analysis of energy cultures in poverty of Abuja, Mumbai and Rio de Janeiro. *Energy Research & Social Science, 72*, 101892.

Delbosc, A. (2017). Delay or forgo? A closer look at youth driver licensing trends in the United States and Australia. *Transportation, 44*(5), 919–926.

Dew, N., Aten, K., & Ferrer, G. (2017). How many admirals does it take to change a light bulb? Organizational innovation, energy efficiency, and the United States Navy's battle over LED lighting. *Energy Research & Social Science, 27*, 57–67.

Filippini, M., & Wekhof, T. (2021). The effect of culture on energy efficient vehicle ownership. *Journal of Environmental Economics and Management, 105*, 102400.

Godbolt, Å. L. (2015). The ethos of energy efficiency: Framing consumer considerations in Norway. *Energy Research & Social Science, 8*, 24–31.

Haerewa, N., Stephenson, J., & Hopkins, D. (2018). Shared mobility in a Māori community Kotuitui, New Zealand. *Journal of Social Sciences, 13*(2), 233–245.

Haustein, S., & Nielsen, T. A. S. (2016). European mobility cultures: A survey-based cluster analysis across 28 European countries. *Journal of Transport Geography, 54*, 173–180.

Hopkins, D. (2017). Destabilising automobility? The emergent mobilities of generation Y. *Ambio, 46*(3), 371–383.

Hopkins, D., & McCarthy, A. (2016). Change trends in urban freight delivery: A qualitative inquiry. *Geoforum, 74*, 158–170.

Hopkins, D., & Stephenson, J. (2016). The replication and reduction of automobility: Findings from Aotearoa New Zealand. *Journal of Transport Geography, 56*, 92–101.

Howden-Chapman, P., Viggers, H., Chapman, R., O'Sullivan, K., Telfar Barnard, L., & Lloyd, B. (2012). Tackling cold housing and fuel poverty in New Zealand: A review of policies, research, and health impacts. *Energy Policy, 49*, 134–142.

Johnson, O. W., Gerber, V., & Muhoza, C. (2019). Gender, culture and energy transitions in rural Africa. *Energy Research & Social Science, 49*, 169–179.

Jürisoo, M., Serenje, N., Mwila, F., Lambe, F., & Osborne, M. (2019). Old habits die hard: Using the energy cultures framework to understand drivers of household-level energy transitions in urban Zambia. *Energy Research and Social Science, 53*, 59–67.

Klaniecki, K., Duse, I. A., Lutz, L. M., Leventon, J., & Abson, D. J. (2020). Applying the energy cultures framework to understand energy systems in the context of rural sustainability transformation. *Energy Policy, 137*, 111092.

Lawson, R., & Williams, J. (2012). Understanding energy cultures. *Proceedings of the annual conference of the Australia and New Zealand academy of marketing (ANZMAC)*. University of New South Wales.

McKague, F., Lawson, R., Scott, M., & Wooliscroft, B. (2016). Understanding the energy consumption choices and coping mechanisms of fuel poor households in New Zealand. *New Zealand Sociology, 31*(1), 106–126.

Rau, H., Moran, P., Manton, R., & Goggins, J. (2020). Changing energy cultures? Household energy use before and after a building energy efficiency retrofit. *Sustainable Cities and Society, 54*, 101983.

Sovacool, B. K., & Griffiths, S. (2020). The cultural barriers to a low-carbon future: A review of six mobility and energy transitions across 28 countries. *Renewable and Sustainable Energy Reviews, 119*, 109569.

Stephenson, J., Hopkins, D., & Doering, A. (2015). Conceptualizing transport transitions: Energy Cultures as an organizing framework. *Wiley Interdisciplinary Reviews: Energy and Environment, 4*(4), 354–364.

Stephenson, J. R., Sovacool, B. K., & Inderberg, T. H. J. (2021). Energy cultures and national decarbonisation pathways. *Renewable and Sustainable Energy Reviews, 137*, 110592.

Tseng, S. H. Y., Higham, J., & Lee, C. (2022). Academic air travel cultures: A framework for reducing academic flying. In K. Bjørkdahl, A. Santiago, & F. Duharte (Eds.), *Academic flying and the means of communication* (pp. 327–353). Palgrave Macmillan.

Urry, J. (2004). The 'system' of automobility. *Theory, Culture & Society, 21*(5), 25–39.

Wells, P., & Xenias, D. (2015). From "Freedom of the open road" to "cocooning": Understanding resistance to change in personal private automobility. *Environmental Innovation and Societal Transitions, 16*, 106–119.

[illegible]

CHAPTER 6

Cultural Change

Introduction

Despite culture's tendency to be relatively stable over time, change does occur. People's adherence to ensembles of thinking, doing and having can alter as they go through life stages, develop new allegiances, take on different beliefs, understandings and aspirations or have new experiences. Membership of culture groups can swell or diminish. Cultural change can be a consequence of changing external influences, such as with the COVID-19 pandemic's normalising of different work practices, or it can be the result of deliberate actions by actors, such as when businesses purposefully adopt more sustainability-oriented cultures. From a sustainability transitions perspective, it is critical to understand the dynamics of cultural change: how cultural transformations are triggered or can be hastened to achieve sustainable outcomes, and how more sustainable cultures can be maintained despite pressures to change.

The cultures framework invites a structured approach to investigating cultural change and cultural transformation. By cultural change, I mean an alteration to at least one aspect of an actor's cultural ensemble that leads to a change in a sustainability outcome. Cultural change can also be considered in terms of fluctuations in the number of actors that share a similar cultural ensemble. By cultural transformation, I refer to more deep-seated change involving adjustments in all aspects of cultural ensembles—motivators, activities and materiality—that have widespread uptake and may also transform structures or cultures at broader scales. Because

J. Stephenson, *Culture and Sustainability*,
https://doi.org/10.1007/978-3-031-25515-1_6

of the system-like qualities of culture, cultural transformation can involve multiple interconnected but lagged changes to a cultural ensemble.

In this chapter, I start by discussing some examples of cultural change. I then move on to illustrating the dynamics of change, looking first at internal change dynamics, and then at change that is stimulated by external influences. I discuss how and why change processes are not always straightforward and can have unintended consequences. The membership of culture groups can change, and I discuss processes by which this occurs. Although cultural actors are individually constrained by their agency, collective cultural change can have repercussions beyond the usual agency limitations. I finish with a discussion of cultural transformation.

Slow and Rapid Cultural Change

Some of the most damaging cultural transformations, when viewed through a sustainability lens, have occurred incrementally over decades or generations as a result of influences that have their roots in colonisation, industrialisation and modernity. In New Zealand, this is visible in the farming sector with the relatively recent normalisation of intensive, high-input agriculture that has been largely responsible for significant declines in water quality (Campbell, 2020). Globally, it can be seen in the way that personalised fossil-fuelled transport, which is a significant causal factor in the climate crisis, became the dominant form of mobility with increasing expectations for larger and more powerful vehicles (Urry, 2004; Vögele et al., 2021).

On the other hand, there are examples of gradual cultural change that has had positive sustainability implications. Many Indigenous cultures, for example, developed cultural knowledge, beliefs and practices to ensure that the natural systems they depended on remained healthy and abundant (Berkes, 1999). In at least some situations, these cultural systems are likely to have arisen from environmental learning, such as discovering the devastating impacts of unsustainable resource use and adapting accordingly, embedding new practices and knowledge within evolving cultural arrangements (Artelle et al., 2018; Wehi et al., 2018). Within the Western world, there are also many examples of the evolution of new beliefs and understandings aligned with sustainability that have gradually become actualised in new practices and products. Environmental movements, for example, which started to build strength from the 1970s as intellectual responses to environmental damage and limits to growth, have shaped new ways of thinking about human relationships with the earth. This has led to pockets and networks of actors who have developed consciously sustainable cultural ensembles, for example among households (Svensson, 2012), farmers (Gosnell et al., 2019) and businesses (Nosratabadi et al., 2019). Although inspiring, these examples are nowhere near the scale of cultural transformations that will be required to rectify global and local sustainability crises, which makes it critically important to understand processes of, and barriers to, wider cultural change.

Several studies using the cultures framework have looked at how culture changes over generations. I have already mentioned the work in Italy comparing the more frugal energy cultures of elderly people from the war generation with the more profligate energy cultures of the baby boomer generation (Bardazzi & Pazienza, 2020). The same researchers also compared the transport-related energy cultures of baby boomers and younger generations in Italy. While people generally use private transport less as they age, this study found that the older cohort had significantly higher private transport use and thus higher transport fuel expenditure than younger generations. The authors interpreted this as an evolving energy culture in younger generations towards more sustainable mobility (Bardazzi & Pazienza, 2018).

Cultures can alter over a period of years, a good example being a study of changing energy cultures in Norway. This work compared households in the 1991–1995 period, when climate change was given little public attention, with 2006–2009 after climate change became a major public concern (Aune et al., 2016). The main change in energy culture was found to be in householders' perceptions of energy, with some evidence also of less consumptive energy practices. In the 1990s, household energy culture emphasised comfort and convenience in everyday life, but by the 2000s households were more concerned about their energy consumption, linking this to climate change. These concerns led to some changes (albeit not radical) in energy consumption, although the dominant expectation was still for a convenient and comfortable lifestyle. The most notable change in energy culture was that households now expressed guilt about their energy consumption because of its links to climate change, and spoke about the difficulties of change. The authors concluded that this may be an early stage in the transformation of household energy cultures, and that despite their resistance to change they may be further reshaped by climate change concerns in the future, especially if supported by targeted policies (Aune et al., 2016).

Cultural change can be even more rapid. National restrictions and rules that were quickly established to combat the spread of COVID-19 had almost immediate repercussions for many aspects of everyday life. A study of impacts on organisational cultures recorded material changes such as closing off open-plan offices with Perspex screens, and the ubiquitous wearing of masks. Water-cooler chat and other social meeting rituals were replaced by Zoom calls. Working from home became normalised. Values and assumptions shifted from creativity and exploration to a focus on safety and resilience (Spicer, 2020). Some of these changes, such as the increased use of videoconferencing and working from home, are likely to continue even after COVID-19 restrictions are lifted and have generally positive implications for sustainability.

For the sustainability transition, cultural change needs to be rapid—we do not have the luxury of slow intergenerational shifts. In order to avoid the worst impacts of climate change and other consequences of living beyond local and planetary limits, we have two to three decades to radically change dominant systems of production and consumption and the cultures that drive these. If the response lags, then chaotic cultural change will likely be forced upon us by unravelling environmental, economic and social systems.

External Influences and Cultural Change

In this section, I discuss how cultural change may be stimulated by a change in external influences, drawing from examples that have used the cultures framework. Most studies have identified at least one external influence that has had a role in impelling change to an element of culture. These external influences may be purposeful (e.g. new policies to support uptake of energy-efficient lightbulbs) or circumstantial (e.g. experiences of climate change impacts leading to new beliefs and expectations about climate change).

Changes to Car-Dependent Cultures

Achieving lower-carbon, more sustainable mobility is a complex quest, as transport involves diverse modes and technologies and operates at multiple scales, with local transport cultures shaped by national and global systems, resource flows, infrastructures and powerful actors. As I have already discussed in Chapter 5, the dominant mobility culture of car dependence is both strongly self-reinforcing (with aligned motivators, activities and materiality) and also locked in with external influences. However, changes to external influences are starting to drive shifts in this dominant transport culture.

In Chapter 5, I discussed a study in New Zealand that compared car-dependent young people with those who had adopted a 'multi-mobility' culture (Hopkins & Stephenson, 2016). The study showed that the culture of the multi-mobility group was supported by number of social, political, physical and financial influences. These included changes to driver licensing, increasing availability of safe cycle lanes and public transport, fewer incentives for car ownership, health messaging about the benefits of active mobility, and concerns about congestion and poor air quality. More general influences included the broadly accepted ideological imperative for action on climate change and the widely shared expectation that young people would travel overseas as a rite of passage (thus prioritising savings over car purchase) (Fig. 6.1).

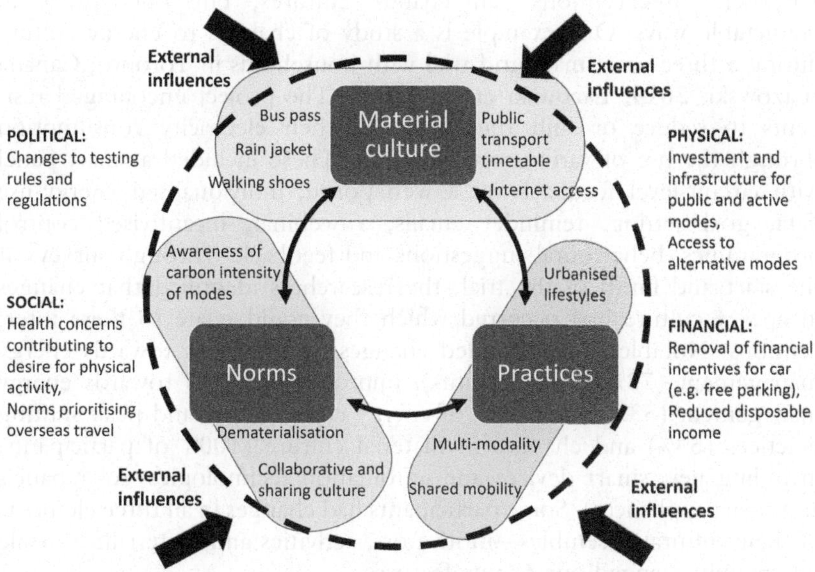

Fig. 6.1 Cultural ensembles of young people with multi-mobility cultures, and external influences that supported these ensembles (Reproduction of Fig. 3 from Hopkins and Stephenson [2016])

This changing context is not confined to New Zealand. In many nations, policy and investment changes have led to new forms of urban design (e.g. the 15-minute neighbourhood), infrastructure (e.g. cycleways) and public transport (e.g. high-speed rail), which make public and shared transport more attractive. Developments in information and communications technologies have enabled a flourishing of new transport types (e.g. mobility as a service) and eased the use of public and shared transport (e.g. real-time arrival times). Technological developments and policy initiatives (such as the increasing cost competitiveness of electric vehicles, targeted subsidies and signalled future bans on imports of fossil fuel cars) are making internal combustion vehicles less attractive. Together, these external influences are resetting the environment that has traditionally reinforced the dominant car culture (Stephenson et al., 2015).

Impacts of Smart Grids on Energy Culture in Canada

Purposeful interventions can change cultures, but not always in predictable ways. One example is a study of changes to energy culture during a three-year smart grid trial with households in Toronto, Canada (Lazowski, 2019; Lazowski et al., 2018). The project encouraged residents to reduce or shift the timing of their electricity consumption through the use of various interventions. These included a smart panel with circuit-level feedback on a web portal, individualised energy-use data, goal setting, reminder emails, a webinar, incentivised control programmes, behavioural suggestions and feedback. Through surveys at the start and finish of the trial, the researchers identified that changes in energy culture had occurred which they could relate to these interventions. Notable shifts included changes in awareness towards energy management (73% of participants), improved attitudes towards energy management (53%), active use of energy conservation and peak shifting practices (53%) and changes in material culture (100% of participants, including new smart devices and automation technologies, solar panels and smart appliances). Some participants had changes in all three elements of their cultural ensembles—motivators, activities and materiality—while others only changed one or two features.

Despite these changes, the overall impact on consumption patterns was relatively minor. The researchers concluded that the adoption of a smarter and more sustainable energy culture within these households was inhibited by a combination of contextual factors (e.g. technical issues, family members with differing energy cultures) and normative lifestyle expectations (e.g. giving priority to comfort and convenience). This indicates how other aspects of cultural ensembles, and other cultural members, can constrain more extensive cultural change and the sustainability benefits that might flow from this. One benefit of this form of analysis was the ability to identify where further adjustments to external influences (e.g. personally tailored information) could help achieve more substantial and lasting changes in culture and thereby in energy consumption.

Interventions for Household Energy Efficiency

The role of purposeful interventions in culture change was also analysed in a study of the effectiveness of two different programmes to encourage households to take actions to improve their energy efficiency (Scott et al., 2016). As already mentioned, much of New Zealand's housing stock is of low thermal quality and this can adversely affect the health and well-being of occupants as well as costing more than necessary to heat. This study trialled different types of energy interventions with householders in three different suburbs in Dunedin, New Zealand. In two suburbs, houses were surveyed by an energy auditor and the occupants were given personalised advice about physical and practice changes to improve efficiency and warmth. In the third suburb, householders were invited to three community energy events that included general advice and practical workshops for greater energy efficiency, including recommendations on simple material changes such as efficient lightbulbs and practice changes such as drawing curtains at night.

The interventions had different effects on the energy culture of the households. Home energy audits were successful in encouraging both behavioural and material changes at an individual level. The energy events, although mainly delivered by professionals, included community members taking part in hands-on activities such as changing lightbulbs, sharing their own stories and tips, and collective discussions on how to save energy based on personal experiences. These built aspirations for change even if some people could not afford to do much due to constraints on personal agency. However, most households undertook at least some changes, and the community events were successful in developing community-wide discussions and awareness of energy efficiency.

The study concluded that the most effective form of intervention may be to offer both programmes: starting with energy education events targeted to the characteristics of the community, and subsequently offering home energy audits. This would enable people to share their thoughts and concerns within the support of their social networks and engender trust in the process, before offering personalised audits, especially where there are cultural barriers to having strangers in one's home.

Loss of Cultural Knowledge in New Zealand

These examples have shown how alterations to external influences, at small and broad scales, may result in changes to cultural ensembles that have improved sustainability outcomes. However, external influences can also erode sustainability-oriented cultural ensembles. As already noted, many Indigenous societies have established effective ways of living within the capacity of their local ecosystems, having developed extensive knowledge over generations about a particular environment and its ecological limits, as well as principles and practices of conduct that will safeguard its health (Berkes & Turner, 2006; Turner & Berkes, 2006). These 'coupled human–environment systems' (Adger et al., 2010) rely on the passing on and enactment of culture knowledge from generation to generation.

A study based on interviews with Māori Kaitiaki (customary environmental guardians) in different regions of New Zealand looked at how cultural knowledge and practices are being lost as a result of environmental degradation (Dick et al., 2012). Modern-day commercial and recreational fishing, along with water pollution from upstream activities had severely affected the abundance and biodiversity of food species in many coastal marine areas. Kaitiaki were distressed by the ecological degradation and equally by the cultural consequences of being unable to catch traditional species. The breakdown in the links between people and their traditional foods had direct implications for the exercise of food gathering, the passing on of cultural knowledge and practices, and ultimately for health, wellbeing and mana. Direct impacts on culture included the severance of the transmission of knowledge and practices specific to species and place; the loss of knowledge of traditional methods of ensuring sustainable harvesting; a reduction in collective events relating to food harvesting, preparation and eating; the erosion of ways in which kinship is maintained (e.g. sharing food with elders, passing on skills to young people); and the inability to access local foods that would usually be given to elderly and infirm community members as a priority. The combined degradation of ecology and culture, and the inability to pass on locally specific knowledge and management practices, undermined the ability of these Indigenous communities to fulfil their culturally defined responsibilities to each other and future generations.

These examples have shown how cultural change can be triggered by influences beyond the control of cultural actors. At a broad scale, these influences may include changes in generally held beliefs or ideologies, changes to environmental conditions, the introduction of new laws and policies, changes in infrastructure and technological developments. Many such influences can be interpreted as alterations in broad-scale structures or cultures. Cultural change can also result from intentional interventions. Examples of successful interventions included sharing new ideas and information, learning skills through hands-on experiences, goal setting, actors feeling that they were part of a community of change, and the provision of new technologies or other material items. As we saw with the Kaitiaki example, changes to external influences can have both direct and indirect implications for culture due to the interdependencies between cultural elements.

Internal Change Dynamics

In Chapter 5, I discussed the linkages between cultural elements within a cultural ensemble, and how these play an important role in cultural stability. If cultural elements change, these same dynamics mean that there can be consequential changes to other features of the cultural ensemble. Using two examples—one relating to businesses and another relating to households in the Pacific—I illustrate the repercussions of change to a single cultural element.

Changing Business Energy Cultures

The first study examined changes that had been made by 142 small businesses in Aotearoa New Zealand to improve their energy efficiency (Walton et al., 2020). Some had been influenced by external factors (e.g. government support schemes) but the important common factor is that these were purposeful changes initiated by the firms themselves. For some, the initial change they made was to material aspects of the firm's energy culture (e.g. the purchase of new energy-related technologies); for others, the initial change was operational (e.g. adopting more efficient routines); and for yet others, the initial change was to the firm's goals and values, often as part of a strategic initiative to reposition the company. In other words, the firms' journeys of cultural change had started with a purposeful alteration to either their motivators, their activities or their materiality.

In many cases, this initial change then had a domino effect on other elements of the firm's energy culture. For example, firms that started with more energy-efficient operational procedures went on to develop other efficiency practices. These changes were often associated with alterations in company expectations such as key performance indicators relating to energy efficiency, and incentives and rewards for efficient staff. Many also undertook subsequent changes to other processes or technologies to continue their efficiency journey. The initial 'trigger point' and consequential changes are illustrated in Fig. 6.2. Costs were a relevant factor in all cases, but cultural characteristics also influenced their choices and change journeys.

Fig. 6.2 A change to energy culture triggered by a change in practices, with consequential effects on materiality and motivators (Reproduction of Fig. 3 from Walton et al. [2020])

For firms that had started with a change to their materiality, the replacement of old technologies with modern ones had sparked them to consider payback periods and cost reductions from energy efficiency. Some subsequently introduced energy monitoring and feedback systems. These changes led to altered norms, knowledge and aspirations about energy. Where a firm's first step had been to revise their values and aspirations (often to build competitive advantage), this led to consequential changes to at least some of the firm's practices and technologies, and in a few cases led to a more sustainable direction for the firm as a whole as it worked to align its operations with its strategic goals (Walton et al., 2020).

As this example shows, the interlinkages between cultural elements means that when there is a change to one element, it can lead to consequential change to other elements. The result can be a domino effect with multiple consequential adjustments to other cultural attributes. Walton et al. (2020) called these 'trigger points' for change.

Lighting in Vanuatu

Another illustration of the trigger and domino effect comes from a study in Vanuatu (Walton et al., 2014). Here, a group of researchers had been asked to evaluate the success of an aid initiative to make portable solar-powered lamps more affordable for communities that had no electricity services. Until then, households had largely been reliant on kerosene lamps, which were expensive, messy, dangerous and caused indoor particulate pollution. Usually households had only one kerosene lamp because of the expense, and it was generally managed by males.

A scheme funded through Australian Aid sought to reduce the price of imported solar lamps at the household level by providing a subsidy to two NGO suppliers to improve their bulk purchasing power. The effect was that good-quality portable solar lamps, which could be recharged for free in the sun every day, were suddenly cheaper and more widely available. The adoption rate was remarkably rapid as households purchased solar lamps, experienced them and spoke with others about the benefits. From a sustainability perspective, they were cheap, safe and easy to operate, and healthier as they eliminated indoor pollution. Within a couple of years, the islands had almost completely transformed from kerosene to solar lighting.

One of the more interesting findings of the evaluation was about the consequential cultural changes from the adoption of solar lamps. Unlike with kerosene lamps, households often acquired more than one solar lamp. These were safe and easy to use, and women and children could now have control of lighting as well as men. Because lighting was available in more than one space, they could undertake activities such as homework and weaving in the evenings. The lamps supported more socialising after dark and were also used for night fishing. Families had less need to work for cash as they no longer had to buy kerosene, so the local economy was more self-sufficient. Alongside these practice changes, people developed new norms and beliefs, such as lighting being for everyone, not just males; an overt aversion to kerosene; and sunlight being more trustworthy as it is made by God. The growing understanding of how to harness sunlight led to new aspirations for other solar appliances, such as small solar panels. These consequential changes to activities, motivators and materiality are illustrated in Fig. 6.3.

This example also illustrates how change to energy culture can have implications for other aspects of everyday life such as food gathering, social processes, livelihoods and education. This considerable cultural shift

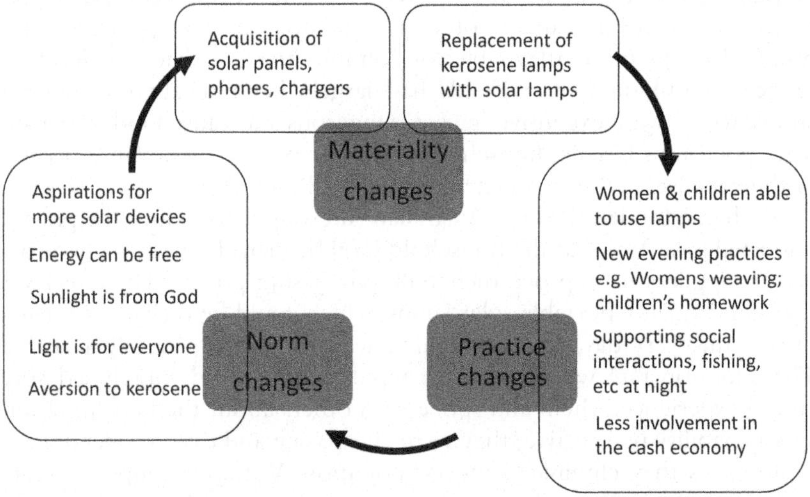

Fig. 6.3 Illustration of consequential changes to cultural ensemble following adoption of solar lamps. Concepts sourced from Walton et al. (2014)

had important sustainability gains for health, livelihoods, equity, education and the environment—although it is not a completely positive story because there will of course be issues with plastic waste as the lamps have a limited life. What it illustrates well, though, is how there can be cascading changes to culture from what seems a small and subtle alteration to a single cultural component.

Another dynamic of change that this example illustrates is serial adoption—how experience with one material change can make another more likely. It is well established in literature on pro-environmental behaviour that prior low-carbon choices are one of the strongest predictors of future low-carbon choices, such that, for example, adopters of solar panels are more likely to consider adopting electric vehicles (Cohen et al., 2019). With the Swiss households discussed in Chapter 5, we saw how the adoption of solar PV was often part of a serial uptake of other more sustainable technologies (heat pumps, thermal solar panels, efficient appliances, etc.), although some non-adopters also possessed these (Bach et al., 2020).

Together, these examples highlight some of the dynamics of cultural change. In the right circumstances, an alteration to one cultural feature can trigger a reset of cultural characteristics, each responding to the other and ultimately resulting in new or differently expressed motivators, activities and materiality. Through the cultures framework, we can envision this process as a series of consequential shifts in actors' cultural characteristics whereby each adjustment is triggered by other changes. The examples also illustrate how the membership of a cultural group can expand through ongoing socialisation and normalisation of those changes across a population.

Cultural change is thus sequential rather than occurring all at once, but this does not necessarily mean that it must be slow. The trick is to identify what is needed to first trigger change and then continue to impel cultural shifts in a more sustainable direction, to avoid reaching a stalling point. This will be greatly dependent on the specific characteristics of any given culture, actors' agency limitations, and the nature and power of external influences.

Culture and the Complexities of Change

Interventions to improve sustainability outcomes rarely take culture into account, and the next three examples show why that can be a mistake.

The first example shows why the intended outcomes of a rural electrification project did not eventuate. In the second example, the benefits of electrification were unequally shared across the community, with women least likely to benefit. The third example shows how cultural analysis prior to the introduction of new technologies can help identify what barriers might be faced, and help in the design of actor-relevant socio-technical systems. Together, the examples show how the interplay between cultural ensembles and external influences can reduce the effectiveness of an intervention and even lead to unintended consequences.

Electrification in Kenya

Studies of electricity use in households are usually concerned about high energy demand and are looking to explain it or seeks ways to reduce consumption. The opposite was the case in a Kenyan study, which sought to find out why rural Kenyan households were consuming so *little* electricity (Tesfamichael et al., 2020). The backstory is that a combined effort by the Kenyan government and industry had led to a large increase in households connected to the grid over the previous 10 years, from 20% to 70%. However, many households were using so little power that the industry was not getting a sufficient return on its investment, and the government was concerned that households were missing out on the welfare benefits of access to clean energy. The difference was most marked in rural settlements, where the average monthly household consumption was 5 kWh compared to 200 kWh in the city of Nairobi. This research examined what was motivating and constraining household electricity consumption in workers' housing in a commercial tea estate. The cultures framework was used because it enabled the researchers to look beyond financial motivations and brought to the fore the multiple influences on energy use, both within the actors' transactional sphere and the wider context.

The householders liked the fact that grid-based electricity enabled them to carry out some desired activities, but they were very cautious about electricity consumption. For the most part, they already had access to other energy sources (kerosene, charcoal, solar and battery-operated appliances) for lighting, cooking and entertainment. Electricity added to their energy options rather than reflecting a wholesale transition from one energy source to another. Where electricity was used, it was predominantly to carry out socially advantageous activities such as night-time study (through better lighting), communication (smartphones), entertainment (music, television) and strengthening social ties (e.g. inviting friends to charge their phones). The arrival of electricity did not radically change the dominant household energy culture, but rather enabled families to enact and reproduce their cultural ensembles in some new ways.

Electricity was useful only to the extent that it assisted with aspirations such as enabling better futures for their children, maintaining ties with their friends and kinship groups, and investing in their longer-term future outside the tea estate. Although cost was a consideration for families, it was not the only influence on their fuel choices. As a result, electricity consumption was very low and other energy-using activities were supported by other fuel types. Their cultural ensembles employed diverse fuels, appliances and practices to reproduce and sustain ways of life that aligned with their aspirations (Tesfamichael et al., 2020).

Culture Change and Gender in Zambia

When a new solar mini-grid was established in rural northern Zambia, it enabled families to partake in modern energy services such as lighting, powering appliances and charging mobile phones. This provided broad benefits for the community, but these were found to be unevenly distributed between men and women (Johnson et al., 2019). Prior to the mini-grid, most households cooked outside or under an open-sided shelter close to the house using firewood burned in traditional cookstoves or three-stone fires. Lighting was typically from burning paraffin, kerosene and candles, with torches used for specific activities. Community economic activity revolved around the use of metabolic energy for fishing, agriculture and harvesting firewood for cooking. The solar minigrid ushered in the potential for households to use power for household activities and livelihoods. Among the interviewed households, the decision to connect was rarely made by women, apart from two who were widows.

By the time of the research, three years after the establishment of the mini-grid, nearly half the households had disconnected. Of those that were still connected, the most ubiquitous use of electricity was for lighting, and in some instances for televisions or radios. When it came to cooking processes, very little had changed as the smart grid did not supply enough electricity for electric cookers, so most households were still reliant on traditional cooking methods. The only fridge in the community was in a medical clinic. Subsistence practices (fishing, agriculture, harvesting firewood, cooking) remained almost unaltered, and there was little disruption to the prevailing norms around roles and responsibilities of men and women. Where difference was identified, it was between connected and disconnected community members, with wealthier business-owning families able to take more advantage of the power compared to poorer subsistence-dependent families.

The energy cultures analysis showed why a material change in energy circumstances does not necessarily empower all members of that community equally. In this case, it mostly advantaged wealthier families. Women in particular missed out on benefits as they already tended to be in a position of less economic and political influence, and their gendered roles in households did not gain from the electrification. The authors concluded that the introduction of new energy systems should be undertaken in parallel with targeted interventions in gender equity and women's empowerment (Johnson et al., 2019).

Adoption of Batteries in the United Kingdom

Where households have PV, battery storage can play an important role in helping reduce the temporal mismatch between electricity generation (when the sun is shining) and highest household need (often in the mornings and evenings when the sun is low or below the horizon). Compared with the relative ubiquity of PV, batteries are not yet widely used at either a household or a community scale, and their future acceptance and use will at least partly hinge on how they fit with public perceptions. This study used the cultures framework to analyse the results of focus groups in Leeds (United Kingdom) with lay members of the public about their views on household and community-scale battery storage. Two of the groups had experience with PV and two did not (Ambrosio-Albalá et al., 2019).

The research sought to identify characteristics of the prevailing energy cultures within which battery technologies might be adopted and used in the future. The researchers were particularly interested in understanding any issues for acceptance and implementation of battery storage. The cultures framework helped the researchers to describe and make links between what participants thought, did and had in energy contexts, and their interests and concerns about future battery technologies, as well as related external influences (Fig. 6.4).

The findings showed that the likelihood of acceptance and adoption of batteries would be shaped by multiple cultural factors including current forms of energy consumption; family expectations; previous experiences; household perceptions of government and the municipal authority; and expectations about the technologies. Battery storage was described as part of an emerging energy culture comprising the material and social fabric of distributed energy systems. While attitudes to domestic-scale battery storage were generally positive, the idea of community-scale battery storage was less favoured, especially if it would involve sharing. On the basis of the study, the authors were able to identify commonly held perspectives (e.g. desires for autonomy, fairness and equity) that would need to be addressed in the design and promulgation of future battery storage schemes (Ambrosio-Albalá et al., 2019).

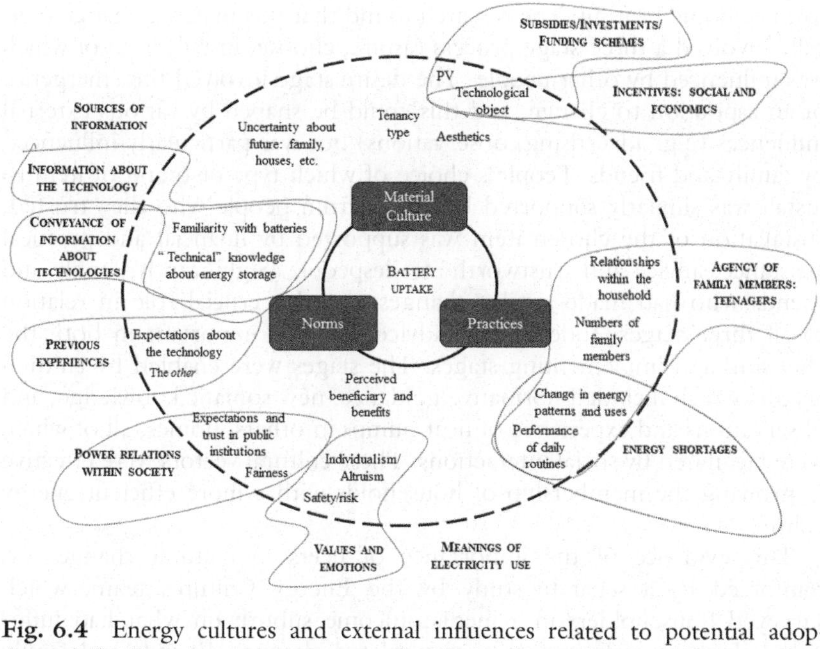

Fig. 6.4 Energy cultures and external influences related to potential adoption of distributed energy storage technologies (Reproduction of Fig. 4 from Ambrosio-Albalá et al. [2019])

Cultural Learning

In Chapter 4, I introduced the idea of cultural vectors to draw attention to the many ways in which culture is learned. Cultural change involves the absorption of new ideas, understandings, norms, practices and other cultural features through vectors such as observations, conversations and hands-on experiences. Part of the dynamic of change is the role of others, especially those we respect and socialise with, in transferring awareness, experiences and aspirations. As we saw with the Vanuatu example, cultural change can become amplified through cultural vectors from a few families to multiple communities, from a singular occurrence to the adoption of similar new cultural features by many actors.

Research in New Zealand by the Energy Cultures research team looked at the role of people's cultural peers—families and friends—in energy-efficient material changes that they had recently made in their homes (Barton et al., 2013). The research drew from focus groups with householders that had recently installed a heat pump—a highly energy-efficient form of home heating. The research found that this material change typically involved a three-stage process (aspire, choose, install) each of which was influenced by other people. The desire stage involved the emergence of an aspiration to change, and this could be shaped by various external influences (e.g. advertising, observations) but was particularly influenced by family and friends. People's choice of which type or brand of item to install was similarly supported by advice from people who they trusted. Installation of the chosen item was supported by financial and practical assistance, advice and trustworthy tradespeople. Significantly, family and friends who had made similar changes played a crucial role in relation to all three stages. Independent advice was also important in both the choosing and implementing stages. The stages were enabled by cultural vectors which included normative guidance, new somatic knowledge, and observations and experience of heat pumps in others' homes, all of which were facilitated by social interactions. These cultural vectors were effective in growing the membership of households with a more efficient energy culture.

This evidence of the importance of peers in cultural change was reinforced by a separate study by the Energy Cultures team, which surveyed householders in a median-income suburb on what had influenced them to make recent energy-related changes. Friends and family members were overwhelmingly reported as the strongest influences.

Personal social networks were found to be 2.9 times more influential than all media together (TV adverts, newspapers, *Consumer* magazine, internet), 3.6 times more influential than all local community groups and 4.4 times more influential than all organisations listed (including the government energy efficiency agency, power companies, council, trades-people and companies supplying energy-related goods). Approximately, 25% reported that their family and friends were the only influence on their energy-related change.

The power of social processes in encouraging cultural change is critically important for sustainability aspirations. We can see this more widely in many fields where change is driven by forms of collective activity and peer alignment. Examples include the expansion in the number of businesses committing to sustainable practices (a fringe activity only a decade ago), the recent growth in individuals committing to veganism or low-meat diets, and the rise in activism on climate-related issues. If we want to hasten cultural shifts towards sustainability, it will be important to pay attention to how people adopt new motivators, activities and materiality through the influence of their peers, and the role of cultural vectors.

Transformational Change

Possibly, the most transformative cultural dynamic is where cultural actors reshape the broader context within which they operate. In the examples so far, cultural actors have been presented as being largely constrained by their agency boundary, with cultural ensembles at least partly shaped by external influences but unable to reciprocally influence those external influences. However, in some instances, the massing of cultural change can enable actors' own influences to extend beyond the agency barrier, to reshape the exogenous forces or structures that usually constrain them. This dynamic is particularly promising for achieving sustainability transitions.

My first example is from New Zealand. Unlike policies in many countries, the New Zealand government has never offered feed-in tariffs or other subsidies to encourage the uptake of solar generation (PV). Indeed, during the second decade of the twenty-first century, almost every influential organisation in the government and the energy sector were aligned against PV, creating a highly unsupportive environment for household adoption of solar generation. As well as the absence of subsidies, electricity retailers effectively penalised households for feeding surplus generation into the grid by only paying them a third of what those households would pay for purchasing power when they were not generating. Electricity sector businesses (generators, retailers and distribution companies) publicly expressed the view that PV had no part to play in New Zealand's future electricity system.

Media messaging from government and sector organisations was also distinctly anti-solar. The government-established Smart Grid Forum argued against solar on the basis that the marginal cost of PV was greater than for wind generation, and that it would cause variability in power supply which would be problematic for the electricity grid. The Parliamentary Commissioner for the Environment argued that PV does not reduce greenhouse gas emissions because it would not help with the most emissions-intensive periods of electricity supply. The Energy Efficiency and Conservation Authority argued that if people had money to invest, it would make more sense to put it into an electric vehicle than PV. The Electricity Authority argued that investing in PV was regressive as ultimately it meant that poorer people would pay more for their electricity. Distribution companies saw it as a problem in that it had the potential to affect power quality and would require them to alter their management systems. The overriding message from these powerful actors was that PV conferred no environmental benefits and (for most households) it would be a lot cheaper to buy power from the electricity grid. This undermined the usual reasons that people give for adopting solar generation—cost savings and environmental benefits.

But despite this unencouraging context, households kept steadily adopting PV, enabled by a several small firms in the solar supply business. Seeking to understand this surprising behaviour, our research found that many households were motivated not by money or environmental concern, but by a lack of trust in electricity companies and a desire to break their dependence on these companies (or to achieve at least partial independence) (Ford et al., 2017). Environmental motivations played a part, but given that New Zealand's electricity was already over 80% renewable this was not a major factor. Households were aware of the costs of installing PV, but savings were not their overriding driver. Even where net present value calculations indicated that their power was more expensive than grid power for some households, they rarely expressed a concern about this.

Through the lens of the cultures framework, PV had a good fit with household aspirations, beliefs and expectations, resulting in adoption rates that would be seen as 'irrational' through a neoclassical economics lens. Over less than a decade, the inexorable increase in solar households eroded the dominant view in the electricity sector that this was a temporary blip of enthusiasts. It shifted to a grudging admittance that uptake was likely to continue, and then to acceptance that PV would play an important role in New Zealand's future electricity systems. In recent years, there have been consequential changes in all parts of the sector: ongoing PV growth is now factored into national energy models and forecasts; large electricity retailers have developed packages specifically for PV owners; and distribution companies have largely accepted that PV will drive unprecedented change in the way lines infrastructure will be managed. The government is now establishing PV on social housing, and large solar farms are starting to become a reality. A flurry of new niche businesses and social enterprises have emerged that offer different models for households to own, lease, share or otherwise benefit from solar generation.

Through a cultural lens, the government and electricity sector's beliefs and expectations were overturned by the widespread shift in energy culture among households. The sector's preference for business-as-usual was derailed by the inexplicable (to them) uptake in new technologies and the implications that this would have for their current business models. As a result, dominant beliefs and norms in the energy sector had to change, with consequential shifts in sector knowledge, practices and materiality, as well as new entities and business models entering the sector. While any individual household would have had no chance of altering the change-averse culture of the dominant regime, it could not resist the collective effect of energy culture change across tens of thousands of households (Ford et al., 2017).

The analysis in Ford et al.'s paper used the cultures framework alongside the multi-level perspective (MLP) on socio-technical transitions (Geels, 2012). It complemented the MLP by providing a way to account for the role of cultural change in transitions, mapping the dynamics of transition from a small group of actors to mainstream, ultimately leading to disruption and transformational culture change among more powerful actors. The role of cultural change in regime shifts can be envisaged as moving from a niche status to destabilising the incumbent regime.

Another example of systemic change from the expansion of a seemingly insignificant cultural shift is the rapid normalisation of divestment from fossil fuels. Divestment was initially an apparently symbolic action that was started by a handful of NGOs and universities from the early 2010s and has now become a core feature of the rapidly growing responsible investment sector. Over 40 trillion dollars had been shifted from the fossil fuel industry at the time of writing, and much of this is now driven by major banks and fund managers. This can be seen as a cultural shift within the finance industry, as it first involved alterations in beliefs about the relevance of investment choice for reducing greenhouse gas emissions, new aspirations by organisations and businesses to change investments, adjustments to internal practices and policies (with concomitant persuasion of key decision-makers that this was important and necessary) and then putting in motion the shifting of large sums of money into (usually) low-carbon alternatives. As more organisations and businesses made the change, more were persuaded to join the divestment movement, thus creating a snowballing culture change with disruptive implications for the fossil fuel sector and positive implications for the renewables sector.

These are just two examples of how cultural shifts that start among traditionally less powerful actors can reshape wider regimes towards more sustainable outcomes. A further example is the effect of new consumer food choices, such as the increase in veganism and vegetarianism. This is leading to changes in food production, in products available in supermarkets and grocery stores, and in food offerings at cafés and restaurants (Boström et al., 2019).

What is notable about all of these examples is how the collectivisation of cultural change among less powerful actors can destabilise more powerful actors (or seen another way, can alter structures and systems that are usually considered beyond their influence). This kind of action would often be analysed through the lens of social movement theory (Gillan, 2020) but this would suggest that the destabilising effects are always intentional. In the New Zealand PV example, adopters of PV were (in the main) simply following their aspirations rather than actively seeking to destabilise the sector, and this may also be the case with many who adopt veganism or vegetarianism. This indicates that while transformative change of regimes can result from purposeful 'movements', it can equally result from the combined effect of cultural shifts by less powerful actors who are just seeking to fulfil their own motivations—a form of social tipping process (Winkelmann et al., 2022). If these motivations happen to align with more sustainable outcomes, then this may force more sustainable adjustments at a regime level. These kinds of changes give me the most hope for a sustainability transition: where isolated cultural shifts that are at odds with the dominant regime mass up and break through the agency barrier to reshape the culture of entire sectors.

For the most part, the examples I have used in this chapter have been about relatively minor cultural changes rather than deep transformations away from unsustainable paradigms, mostly because the latter are not yet well studied through a cultural lens. More research is needed to better understand processes of cultural change, particularly where they lead to fundamental shifts in unsustainable ideologies and paradigms. I am inspired by small nations that have deliberately adjusted away from growth models to other measures of success, such as 'sufficiency thinking' in Thailand (Avery, 2020) and 'gross national happiness' in Bhutan (Munro, 2016). I am also inspired by groups of people within dominant unsustainable cultures who are demonstrating completely different ways of thinking, doing and having, such as regenerative and community-enhancing forms of agriculture (Bisht & Rana, 2020; IPES-Food, 2018; Sumner et al., 2011) and community energy projects that deliver renewable and affordable energy (Fuller, 2017; Parag & Ainspan, 2019; Watts, 2018). To achieve the sustainability transition, more needs to be known about the conditions under which cultural actors, at any scale, are willing and able to make fundamental changes to their motivations, activities and materialities.

Conclusion

Cultural change is often a slow, incremental process, but faced with the sustainability crisis it is imperative that it occurs rapidly and infectiously. Research, some of it using the cultures framework, is starting to reveal processes of cultural change as they relate to sustainability issues. The examples I have described show how cultural change may be initiated, and the subsequent dynamics of change. Starting points include one or multiple external influences, and/or internal (actor-driven) adjustments of a feature of their cultural ensemble. Vectors such as observations, bodily experiences and social interactions can support the adoption of new cultural features. Once cultural change is initiated, there may be consequential changes to other cultural elements, which in the right circumstances can lead to cultural transformation. Where cultural change becomes widespread among a population, it may destabilise unsustainable regimes that would usually be beyond the power of cultural actors to influence.

However, cultural change is not always a straightforward process. As the examples show, there is potential for cultural resistance, unintended consequences and inequitable impacts. Attempts to purposefully change cultures for more sustainable outcomes may have more benefits than intended (as with the Vanuatu example), fewer benefits than intended (as with the Kenya example) or highly unequal benefits (as with the Zambia example). In designing interventions for change, it is therefore important to understand the relevant cultural characteristics of the subject population. The next chapter outlines how to use the cultures framework to support the development of culturally relevant policies and other interventions for change.

References

Adger, W. N., Brown, K., & Conway, D. (2010). Progress in global environmental change. *Global Environmental Change, 20*(4), 547–549.

Ambrosio-Albalá, P., Upham, P., & Bale, C. S. E. (2019). Purely ornamental? Public perceptions of distributed energy storage in the United Kingdom. *Energy Research & Social Science, 48*, 139–150.

Artelle, K. A., Stephenson, J., Bragg, C., Housty, J. A., Housty, W. G., Kawharu, M., & Turner, N. J. (2018). Values-led management: The guidance of place-based values in environmental relationships of the past, present, and future. *Ecology and Society, 23*(3), 35.

Aune, M., Godbolt, Å. L., Sørensen, K. H., Ryghaug, M., Karlstrøm, H., & Næss, R. (2016). Concerned consumption. Global warming changing household domestication of energy. *Energy Policy, 98*, 290–297.

Avery, G. C. (Ed.). (2020). *Sufficiency thinking: Thailand's gift to an unsustainable world*. Routledge.

Bach, L., Hopkins, D., & Stephenson, J. (2020). Solar electricity cultures: Household adoption dynamics and energy policy in Switzerland. *Energy Research and Social Science, 63*, 101395.

Bardazzi, R., & Pazienza, M. G. (2018). Ageing and private transport fuel expenditure: Do generations matter? *Energy Policy, 117*, 396–405.

Bardazzi, R., & Pazienza, M. G. (2020). When I was your age: Generational effects on long-run residential energy consumption in Italy. *Energy Research & Social Science, 70*, 101611.

Barton, B., Blackwell, S., Carrington, G., Ford, R., Lawson, R., Stephenson, J., Thorsnes, P., & Williams, J. (2013). *Energy cultures: Implication for policymakers*. Centre for Sustainability, University of Otago.

Berkes, F. (1999). *Sacred ecology: Traditional ecological knowledge and resource management*. Taylor & Francis.

Berkes, F., & Turner, N. J. (2006). Knowledge, learning and the evolution of conservation practice for social-ecological system resilience. *Human Ecology, 34*(4), 479–494.

Bisht, I. S., & Rana, J. C. (2020). Reviving the spiritual roots of agriculture for sustainability in farming and food systems: Lessons learned from peasant farming of Uttarakhand Hills in North-western India. *American Journal of Food and Nutrition, 8*(1), 12–15.

Boström, M., Micheletti, M., & Oosterveer, P. (2019). *The Oxford handbook of political consumerism*. Oxford University Press.

Campbell, H. (2020). *Farming inside invisible worlds: Modernist agriculture and its consequences*. Bloomsbury Academic.

Cohen, J., Azarova, V., Kollmann, A., & Reichl, J. (2019). Q-complementarity in household adoption of photovoltaics and electricity-intensive goods: The case of electric vehicles. *Energy Economics, 83*, 567–577.

Dick, J., Stephenson, J., Kirikiri, R., Moller, H., & Turner, R. (2012). Listening to the Kaitiaki: Consequences of the loss of abundance and biodiversity of coastal ecosystems in Aotearoa New Zealand. *MAI Journal, 1*(2), 117–130.

Ford, R., Walton, S., Stephenson, J., Rees, D., Scott, M., King, G., Williams, J., & Wooliscroft, B. (2017). Emerging energy transitions: PV uptake beyond subsidies. *Technological Forecasting and Social Change, 117*, 138–150.

Fuller, S. (2017). Embedding energy transitions in the community. In S. Bouzarovski, M. J. Pasqualetti, & C. B. Vanesa (Eds.), *The Routledge research companion to energy geographies* (pp. 265–276). Routledge.

Geels, F. W. (2012). A socio-technical analysis of low-carbon transitions: Introducing the multi-level perspective into transport studies. *Journal of Transport Geography, 24*, 471–482.

Gillan, K. (2020). Temporality in social movement theory: Vectors and events in the neoliberal timescape. *Social Movement Studies, 19*(5–6), 516–536.

Gosnell, H., Gill, N., & Voyer, M. (2019). Transformational adaptation on the farm: Processes of change and persistence in transitions to 'climate-smart' regenerative agriculture. *Global Environmental Change, 59*, 101965.

Hopkins, D., & Stephenson, J. (2016). The replication and reduction of automobility: Findings from Aotearoa New Zealand. *Journal of Transport Geography, 56*, 92–101.

IPES-Food. (2018). *Breaking away from industrial food and farming systems: Seven case studies of agroecological transition*. Food and Agriculture Organisation.

Johnson, O. W., Gerber, V., & Muhoza, C. (2019). Gender, culture and energy transitions in rural Africa. *Energy Research & Social Science, 49*, 169–179.

Lazowski, B. (2019). *Engaging beyond the meter: Encouraging residential energy management using smart grid tools*. University of Waterloo.

Lazowski, B., Parker, P., & Rowlands, I. H. (2018). Towards a smart and sustainable residential energy culture: Assessing participant feedback from a long-term smart grid pilot project. *Energy, Sustainability and Society, 8*(1), 1–21.

Munro, L. T. (2016). Where did Bhutan's gross national happiness come from? The origins of an invented tradition. *Asian Affairs, 47*(1), 71–92.

Nosratabadi, S., Mosavi, A., Shamshirband, S., Zavadskas, E. K., Rakotonirainy, A., & Chau, K. W. (2019). Sustainable business models: A review. *Sustainability (Switzerland), 11*(6), 1–30.

Parag, Y., & Ainspan, M. (2019). Sustainable microgrids: Economic, environmental and social costs and benefits of microgrid deployment. *Energy for Sustainable Development, 52*, 72–81.

Scott, M. G., McCarthy, A., Ford, R., Stephenson, J., & Gorrie, S. (2016). Evaluating the impact of energy interventions: Home audits vs. community events. *Energy Efficiency, 9*(6), 1221–1240.

Spicer, A. (2020). Organizational culture and COVID-19. *Journal of Management Studies, 57*(8), 1737–1740.

Stephenson, J., Hopkins, D., & Doering, A. (2015). Conceptualizing transport transitions: Energy cultures as an organizing framework. *Wiley Interdisciplinary Reviews: Energy and Environment, 4*(4), 354–364.

Sumner, J., Mair, H., & Nelson, E. (2011). Putting the culture back into agriculture: Civic engagement, community and the celebration of local food. *Urban Agriculture: Diverse Activities and Benefits for City Society, 5903*, 54–61.

Svensson, E. (2012). Achieving sustainable lifestyles? Socio-cultural dispositions, collective action and material culture as problems and possibilities. *Local Environment: THe International Journal of Justice and Sustainability, 17*(3), 369–386.

Tesfamichael, M., Bastille, C., & Leach, M. (2020). Eager to connect, cautious to consume: An integrated view of the drivers and motivations for electricity consumption among rural households in Kenya. *Energy Research & Social Science, 63*, 101394.

Turner, N. J., & Berkes, F. (2006). Developing resource management and conservation. *Human Ecology, 34*(4), 475–478.

Urry, J. (2004). The 'system' of automobility. *Theory, Culture & Society, 21*(5), 25–39.

Vögele, S., Hanna, L., Otte, S., & Rübbelke, D. (2021). Why the trend towards gas-guzzlers? A closer look at the complex effects of social norms on German car buyers. *Energy Research and Social Science, 72*(3), 101840.

Walton, S., Doering, A., Gabriel, C.-A., & Ford, R. (2014). *Energy transitions: Lighting in Vanuatu*. Project report for Australian Aid.

Walton, S., Zhang, A., & O'Kane, C. (2020). Energy eco-innovations for sustainable development: Exploring organizational strategic capabilities through an energy cultures framework. *Business Strategy and the Environment, 29*(3), 812–826.

Watts, L. (2018). *Energy at the end of the world: An Orkney Islands saga*. MIT Press.

Wehi, P. M., Cox, M. P., Roa, T., & Whaanga, H. (2018). Human perceptions of megafaunal extinction events revealed by linguistic analysis of indigenous oral traditions. *Human Ecology, 46*(4), 461–470.

Winkelmann, R., Donges, J. F., Smith, E. K., Milkoreit, M., Eder, C., Heitzig, J., Katsanidou, A., Wiedermann, M., Wunderling, N., & Lenton, T. M. (2022). Social tipping processes towards climate action: A conceptual framework. *Ecological Economics, 192*, 107242.

CHAPTER 7

Using the Cultures Framework for Policy Analysis

Introduction

One way to help achieve a more sustainable future is through the development of new policies or other interventions designed to change unsustainable cultural ensembles or maintain sustainable ones. By policy, I refer to intentional actions by governance agencies to achieve different outcomes from the status quo. Policy interventions typically include regulations, incentives, subsidies, information campaigns, nudge techniques and social marketing that seek to change behaviour or encourage the uptake of new technologies.

Cultural analysis has much to offer policy development. Almost every study using the cultures framework has concluded by discussing the policy implications of this form of analysis. I have also used it myself, along with my research teams, to develop policy advice from two major research programmes in New Zealand. This gives me confidence in proposing the cultures framework as a helpful analytical approach for policy development. For those readers who have come straight to this chapter, the cultures framework is described in detail in Chapter 4 but in brief it presents culture as comprising three core elements—shared or common ways of thinking and knowing (motivators), doing (activities) and having (materialities). These elements and their interactions comprise the cultural ensemble of any given group of actors (households, businesses, etc.).

J. Stephenson, *Culture and Sustainability*,
https://doi.org/10.1007/978-3-031-25515-1_7

A cultural analysis can show, for example, why the adoption of new technologies is rarely as straightforward as might be suggested by a cost–benefit analysis or consumers' stated preferences. As the previous two chapters have illustrated, a cultural perspective shows how any decision regarding a new technology is affected by actors' cultural ensembles—including the ideas and experiences deriving from people's past (as with the battery example in the UK), and the current cultural ensembles into which the new technologies would need to fit (as with the PV example in Switzerland). Decisions are influenced by other cultural factors such as internally conflicting motivators (as with Norwegian households and energy efficiency), practices that align with current technologies (as with the US Navy and with timber companies in New Zealand), and shared beliefs that dismiss alternatives (as with car-centric young people in New Zealand). Decisions are also shaped by wider prevailing cultures and structures that support the status quo (as with freight drivers and households in fuel poverty in New Zealand).

In this chapter, I describe examples of policy advice that have been developed from research using the cultures framework, mainly relating to energy and mobility issues. I discuss examples of integrated policy development that have used the cultures framework at every stage from data gathering to the design of interventions. I propose a series of steps that can be followed to develop policy using the framework and outline how the framework can be used to underpin policy evaluation. I start by discussing the extent to which culture is usually a consideration in policy development.

Policy for a Sustainable Future

Policy development is generally still strongly shaped by economic theories. Neoclassical economics assumes humans act rationally to maximise their utility. Through this lens, many human behaviours are mystifying. Why don't people's actions align with their professed beliefs and attitudes? Why don't people act to optimise their financial situation, or act in ways that align with their knowledge of the consequences of their actions? Behavioural economics, a derivative of neoclassical theories, seeks to explain such conundrums by introducing the idea that people's decisions can be shaped by their heuristics (mental decision tools) and systematic biases, as well as the context in which a decision is made (Reisch, 2017). Behavioural economics has been strongly influential in policy development over the past decade, particularly nudge theory, which focuses on making the 'right' decision the least taxing one (Thaler & Sunstein, 2008). Through these lenses, policies and other interventions that are intended to change human behaviour assume that humans are self-interested, lazy and economically driven. Furthermore, responsibility for the sustainability crisis is predominantly presented as an individualised problem—that it is caused by people making poor decisions in their everyday lives. Organisations, powerful actors, ideologies and institutions that maintain the status quo are largely exempt from consideration.

Accordingly, mainstream policy development focuses on influencing individuals' choices. It assumes that people can be persuaded to do things differently if the price is right, and if they have sufficient information, and that the 'right' decision is less effort than the wrong one. Human decision-making and behaviours are undoubtedly shaped by economic optimisation, knowledge and predictable psychological reactions ... to a point. What is missing is the fact that people rarely act as entirely independent beings—our responses to external influences such as policies are shaped by the cultures of which we are part. As the studies discussed in Chapters 5 and 6 have shown, people can be locked into relatively unchanging cultural ensembles due to the positive dynamics between their motivators, activities and materialities. External influences other than the new policy may also be tending to reinforce that cultural ensemble. To take an obvious example, people who are locked into car cultures through good roads and plenty of carparks are unlikely to be strongly influenced to change to cycling by a line on the road that marks a cycle lane.

Policy development is also often aimed generically at a population, rather than targeted to groups with different characteristics. Shortcomings with one-size-fits-all policies are revealed by a review of household energy efficiency interventions in developed nations which found that 90% of interventions took a general population approach, but that those interventions that featured segmentation, targeting and tailoring were more effective (Russell-Bennett et al., 2019). Another study of the uptake of low-carbon technologies and behaviours concluded that 'policymakers of all types ought to move from a focus always on individuals to a recognition that group-based and collective phenomena—such as culture—shape and influence aspirations, capabilities, and agency for low-carbon transitions' (Sovacool & Griffiths, 2020: 9).

More fundamentally, sustainability transitions will not be achieved through incremental approaches, but instead will require rapid transformational changes at multiple scales beyond individuals. Scholars of socio-technical transitions, for example, understand them as involving transformational changes across multiple actors and multiple dimensions including technologies, markets, policies, industries and civil society (Geels, 2012). Transition processes are sometimes framed in techno-economic terms, focusing on the role of resource flows and technological developments along with evolutions in markets, or in political terms, focusing on the roles of policies and political actors (Cherp et al., 2018). Transitions theories have made significant strides in illuminating aspects of the complex changes involved in achieving a sustainable future, but I believe they are hamstrung by not adequately accounting for the workings of culture in both resisting and driving change.

Culture is most visible in socio-technical framings of transitions, but is considered as a minor factor if at all. Where culture is included in analysis, it is equated to symbolic meaning (Geels, 2002), visions and values (Geels, 2005), belief systems (Geels et al., 2007), discourse and public opinion (Geels, 2011), and with collective sense-making (Geels & Verhees, 2011). While these are all important aspects of culture, they confine its interpretation to the cognitive realm rather than its entangled existence with the physical and active dimensions of culture. They also fail to account for how culture can operate structurally to maintain unsustainable regimes. I believe policy-making needs to grapple with how to support the transformation of existing regimes, and that this will require a better appreciation of the influence of culture in supporting the status quo, as well as its transformational potential.

The examples in Chapters 5 and 6 show why an understanding of socio-cultural processes is critical for the development of sustainability-oriented policy and/or other interventions for change. They illustrate the constraining influence of actors' existing cultural ensembles, whereby an idealised choice may have a poor fit with what people or organisations already have, think and do, as with batteries in UK households. The examples show how policies may fail where multiple external influences reinforce an existing cultural ensemble, so that a single policy signal may play only a minor dissenting part in a loud chorus of support, as with automobility. They show how actors' choices are constrained by the limited scope of their agency, as with households in energy poverty. Policy approaches that fail to account for culture may fail to anticipate consequences, such as for gender equity in Zambia, may have regressive impacts as with slum rehabilitation housing or may result in more extreme change than anticipated, as with solar lighting in Vanuatu. The examples also show how cultural change can unexpectedly occur despite a non-supportive policy environment, as with PV uptake in New Zealand, or can reshape the policy environment, as with the growth and mainstreaming of fossil fuel divestment. By becoming more aware of the transformational possibilities of culture as well as the way culture can act as a constraint and a structure, a cultural analysis can help avoid 'unjust, hegemonic, or narrow narratives of development and implementation' (Sovacool & Griffiths, 2020: 9) and can open up possibilities for policy actions that support transformational change.

A focus on culture rather than individual behaviour invites a different policy development approach. Behaviour invites policymakers to consider changing what people do. Culture invites a focus on why and how they do it within their social context. A cultural approach recognises that people are neither completely individualistic, autonomous and anarchistic, nor completely socially homogenous, socially fettered and socially dependent. They have some freedom to make choices, but the scope of their agency can be limited. People's responses to policy can be constrained both by their existing cultural ensemble and by the multitude of external influences that reinforce that ensemble. A cultural approach can also help reveal heterogeneity across populations, so that policies can be better targeted to groups of actors who have similar motivators, activities and materialities. Additionally, culture opens the door to analysis of cultural processes at multiple scales, including considering the roles of powerful institutions and actors, and the role of culture within policy agencies themselves.

The implications for policy of many of these ideas are already articulated elsewhere, such as the interplay between agency and structure (Sandfort & Moulton, 2020), the adoption of innovations (Berry & Berry, 2018), the constraining effects of practices (Sahakian & Wilhite, 2014; Shove, 2014) and the governance of socio-technical systems (Borrás & Edler, 2020; Geels et al., 2017). The cultures framework provides an alternate perspective that can complement these approaches (e.g. Ford et al., 2017). It can also be used as a framing for policy analysis in its own right, which is the topic of the rest of this chapter.

Applications of the Cultures Framework to Specific Policy Questions

The cultures framework has been used to inform policy recommendations over a wide range of topics. The following examples illustrate how the framework can be used and the kinds of policy recommendations that have emerged from a cultural analysis.

Improving Driving Efficiency

One of the central hypotheses in mainstream policy-making is that people fail to make the right choices because they have insufficient knowledge. A study using the cultures framework reveals one of the flaws of this assumption (Scott & Lawson, 2018).

Efficient driving is a zero-cost way to reduce fuel use by 10–20% without significantly increasing trip times. The study authors carried out focus groups with drivers, asking about their driving practices and their knowledge of efficient driving. In every case, the participants knew cognitively what was involved in driving efficiently—they were well aware of the techniques involved—but said that they rarely did so in practice. Clearly, information campaigns would not make a difference here. Relating this to my discussion of cultural learning in Chapter 4, participants appeared to have semantic knowledge but not bodily knowledge relating to driving efficiency. Their driving patterns were dominated by the inefficient driving skills that they had learned and embodied, not their cognitive understandings of efficient driving. The research also showed that participants rarely connected their driving practices to carbon emissions. If they did drive efficiently, it was usually linked to cost-saving rather than environmental concerns.

The researchers proposed three types of intervention, all seeking to change aspects of driving culture. Relating to material interventions, they considered that some drivers would respond well to in-car smart feedback devices that highlight the real-time fossil fuel consumption or emissions. The second proposal was to develop a link between knowledge and outcomes through social marketing messages that showed how driving practices have a significant effect on carbon emissions and therefore on climate change. Third, they recommended that mastering efficient driving skills should be one of the requirements of gaining a driver's licence, and that advanced classes in driving efficiently should also be offered (Scott &

Lawson, 2018). Relating this to my discussion in Chapter 4 about the different pathways through which culture is learnt, this would build up efficiency routines as bodily knowledge, complementing drivers' semantic knowledge of efficiency but being able to be drawn on without cognitive effort.

Reducing Energy Injustices

A cultural analysis can also help avoid inequitable or unjust outcomes from interventions. This study of energy injustices in slum rehabilitation housing developments has already been outlined in Chapter 5 (Debnath et al., 2021). One of its intentions was to support the design of just policy, with a particular focus on Sustainable Development Goals 3 (good health and wellbeing), 7 (clean and affordable energy) and 11 (sustainable cities and society). The analysis revealed interdependencies between the design of the rehabilitation housing, the resulting energy cultures of the communities and the distributive injustices arising from this. The qualities of the respective built environments were found to be a crucial distributive justice factor, along with affordability and quality of appliances.

Policy recommendations from this investigation of energy cultures covered three types of intervention. First, the design of transitional housing should support the cultural identities of the relocated communities, as the evidence showed that the forms of rehabilitation housing altered their energy cultures in ways that created injustices. Second, the governance of energy utilities and their interactions with households (administrative lags, irregular billing cycles, low power quality) was another cause of injustice for low-income households and required policy intervention. The third policy consequence was to support households to switch to cleaner fuels, as the households currently dealt with their energy insecurities by utilising multiple fuels including firewood, kerosene and LPG as well as electricity, with resulting health implications from their use indoors. The research also found that community-driven initiatives (e.g. refrigeration sharing, rooftop solar) could help alleviate some inequities, as could the availability of affordable repair and maintenance shops for appliances.

From a cultures framework perspective, the first two policy recommendations are addressing structure-like external influences on energy injustice—one relating to built infrastructure and the other to business practices. Only the third policy would directly interact with households, yet all are critical to overcoming energy injustices and attaining the relevant Sustainable Development Goals.

Policy to Address Energy Poverty

Several studies have used the cultures framework to help understand the multiple dimensions of energy poverty, and from this have recommended policy interventions that would assist in achieving greater energy wellbeing. A study in southern Chile examined the energy cultures of households in energy poverty, and among other things it identified the key role of firewood in heating and cooking. There was also a wider culture beyond individual households, involving the livelihoods of small local firewood retailers. Current policies focused on providing access to electricity and did not account for the implications of many households being largely reliant on firewood, nor the indoor air pollution arising from this. Policies also failed to account for the historically contingent forms of housing or the implications of local climatic conditions. The authors called for a context-sensitive approach to policy that accounted for local energy cultures. This would include participatory decision-making to help design policies that recognise and integrate citizens' energy culture and to use this to underpin the design of culturally appropriate energy efficiency programmes (Cortés & Amigo, 2022).

Another household-based study used the cultures framework to gain a more comprehensive understanding of energy poverty beyond the usual narrow triad of income, housing and energy appliances (McKague et al., 2019). The analysis revealed the heterogeneity of experiences of energy poverty. A household's vulnerability to energy poverty is shaped by a wide range of factors including their cultural ensemble, agency and external influences, and is experienced in many different ways. Interventions are unlikely to be effective if they just focus on one aspect of the problem, such as winter fuel payments that assume the main problem is lack of income. The study findings suggested, for example, that people who spend extended amounts of time in the home, such as the elderly and families with young children, would benefit from direct interventions for material improvements, such as subsidies for heating and insulation. Those who would benefit by using energy more efficiently could receive customised home energy advice, for example regarding energy-saving practices and small cost-effective investments. Better understanding of the energy cultures of households can help to design targeted interventions for groups of users who may be demographically similar but experience energy poverty in diverse ways.

Reducing Peak Demand

The variability of electricity demand is an important sustainability issue because peaks in demand in electricity networks often need to be met with fossil-fuelled power generation. In many nations, morning and evening peaks in demand are mostly driven by activities in households, so understanding these activities is an important first step in reducing demand peaks and thus reducing carbon emissions. Studies on the causes of household peak demand are typically complex and expensive to undertake and require detailed energy consumption data, and most such studies to date have been undertaken in developed nations. In many developing nations, the electricity grid is quite vulnerable to brown-outs or black-outs when demand peaks cannot be met by supplies, and understanding the drivers of demand peaks would assist in reducing this vulnerability. However, detailed consumption data is rarely available due to the absence of smart meters, so the costs of peak demand studies may be prohibitive.

A relatively low-cost approach that did not require half-hourly consumption data from household meters was developed in Bangladesh based on the cultures framework (Khan, 2021). Through surveys with householders, Khan collected data on factors that have the potential to affect the timing of electricity demand, including material aspects (e.g. appliances, cooking fuels, house size), practices (appliance use, energy-saving behaviours), motivators (knowledge, aspirations, attitudes) and demographic characteristics. Through analysis using the demand profile of appliances and reported patterns of practices, he found that some households had markedly higher consumption than others during periods of peak demand on the overall grid. The two biggest contributors to peak demand in households were rice cookers and the use of air conditioning. The 'peaky' households were mostly larger, had more electrical appliances, were owner-occupied, and had higher incomes and more children. These households were more concerned about the environment but less interested in reducing consumption to save money than non-peaky households. Khan concluded that the best policy response would be to introduce efficiency standards for rice cookers and air conditioners to reduce demand, since given the daily routines of households it would be very difficult to shift the timing of demand (Khan, 2021).

Domestic Water Use

The cultures framework has also been used as an integrative model across multiple data sets to support better policy development (Manouseli et al., 2018). In the UK, domestic water supplies are challenged by climate change, and there is a need to develop evidence-based drought scenario models for the purpose of water management. There was limited evidence available on the factors driving domestic water use in drought and non-drought circumstances. This study used the cultures framework to underpin integration of the available evidence on interactions between social norms, practices and materiality. The framework enabled them to link data on motivators (such as comfort, cleanliness, garden 'greenness'), practices (when and how water was used) and materiality (water-using devices, water-saving devices, metering). This revealed the existence of distinctive water cultures across the population with different levels of water demand. From a policy perspective, the authors suggested that these cultures would have different responses to price changes, media messaging, new technologies or drought management interventions, so policies would ideally be targeted to the different water cultures. The cultures framework thus acted in two ways to assist with the research—as an integrating frame for multiple data sets, and to identify the heterogeneity of water cultures as a basis for policy design.

Energy Technologies and Collective Action

A cross-national European study compared countries' policies that supported collective action for low-carbon energy transitions (Carrus et al., 2019). Case studies were undertaken in six countries and identified similarities and differences in energy cultures in respect of the uptake of electric mobility, smart energy technologies and energy-efficient buildings. In most cases, adoption was associated with strong environmental motivations, a strong social support system and clear financial incentives. Common barriers to change were legal uncertainties and bureaucratic burdens for individuals and groups wanting to start initiatives.

Success in implementing collective action was related to factors within actors' cultures as well as to country-specific external influences.

The study confirmed the importance of designing policy that recognises cultural heterogeneity. Recommendations included addressing the policy culture (e.g. harmonising different regulatory frames, congruence between national, regional and local policies) and resetting the policy context so that collective initiatives could flourish (e.g. active support of initiatives, easy access to funding). Some recommendations were country specific, and others were generic policy recommendations to support collective approaches by private citizens and businesses.

These examples reveal how cultural analysis can assist in more effective policy development. Policy developed with a lack of understanding of culture may be unsuccessful and even regressive. Financial or information-based policies can only influence certain aspects of cultural ensembles and cultural learning, and may be ineffective due to resistance derived from other aspects of culture. The examples reveal how a cultural approach differs from a demographic approach to policy design, in that it accounts for what people think, have and do. They show the importance of designing policy to suit existing cultures, whether that is at a national scale or focused on locally specific cultures that have a particular relationship with place-based resources. Even well-intentioned policy interventions can bring about or worsen injustices if they fail to account for culture. The examples show the importance of supporting and enabling initiatives that are already working towards sustainable outcomes. Finally, they draw attention to policy culture itself, and the importance of congruence between regulations and between national, regional and local policies in order to support sustainability transitions.

Undertaking a Programme of Policy-Relevant Research

Policy development needs to be supported by data, which will often need to be derived from research. In this section, I describe how the Energy Cultures research teams used the cultures framework to underpin the development of policy advice in two tranches—a three-year research programme ('Energy Cultures 1') culminating in a policy report in 2013, followed by a four-year programme ('Energy Cultures 2') culminating in a policy report in 2016. This is not to say that multi-year research programmes are always needed; the purpose here is to explain how our teams undertook these programmes of policy-relevant cultural research.

Energy Cultures 1

Our 2013 policy report, based on three years' research on barriers to energy efficiency, focused on research-informed interventions to enhance household energy efficiency (Barton et al., 2013). It may be useful to refer back to the figures in Chapter 4, especially Figs. 4.4 and 4.6, to visualise the following discussion.

The research underpinning the policy advice examined different aspects of relevant cultural ensembles and their dynamics. Some of this work has already been individually reported in earlier chapters, but it is important to show the scope of the programme as a whole. Householder interviews on the links between their values and energy-efficient behaviours explicitly looked at the motivator-activities dynamic of the cultural ensemble. Choice modelling explored people's preferences for various attributes of heating and hot water systems, explicitly looking at the motivator-materiality dynamic, and identified clusters based on the dominant preferred attributes. A national survey of households gathered data on motivators, activities and materiality of a representative sample of households. This enabled (among other things) the identification of four main clusters of household energy cultures differentiated by cultural features relating to energy efficiency. Focus groups and surveys explored householders' perspectives on what and who stimulated them to make an efficiency change, identifying cultural vectors as well as internal and external drivers of successful change. Other interviews looked at barriers to a particular efficiency change. A community-based trial studied the different outcomes from two different forms of information-based interventions with households. Studies of law, policy and performance standards identified external influences on the efficiency of household practices and materiality.

The cultures framework thus supported an interdisciplinary, integrative approach to data acquisition and analysis. A range of different research methodologies produced qualitative and quantitative data on the energy culture of New Zealand households, and the findings were then integrated to develop policy-relevant insights. The policy recommendations included proposed changes to law, standards, subsidies and policies for the population in general to make culture change easier; advice targeted to the different clusters of energy cultures; and interventions crafted to better support the journeys that households undertake in adopting new practices and material items (Barton et al., 2013). The specifics of the advice can be found in the policy report; however this is not necessarily transferable to other situations given that it is, after all, a cultural study, particular to time, people and place. What *is* transferable is the approach outlined here, of using the framework to underpin the design of multiple pieces of research that explore different variables and their dynamics, which are then triangulated and integrated to develop policy recommendations.

Energy Cultures 2

A subsequent programme of research used the cultures framework to explore two topics that had been put forward by our primary funder, a government agency (Stephenson et al., 2016). One question was about how to leverage energy savings in homes, small businesses and transport, and the other was about ways to encourage consumers to adopt energy-efficient transport options, and how to encourage markets to deliver them.

Unlike the previous research programme, which focused on household energy efficiency, the research team in this case was asked to deliver on several outcomes (energy savings, energy efficiency, technology adoption) across several sectors (households, businesses, transport users, markets and policy agencies themselves). Accordingly, rather than a single integrated research design, we applied the cultures framework to designing a number of parallel research projects working in each of the sectors. Similar to Energy Cultures 1, this involved researchers from multiple disciplines applying a range of methodologies. The framework provided a common 'language' for all researchers and was a connector across all of the projects, enabling us to share insights about cultural dynamics even where instrumental findings were sector specific.

The interdisciplinary research underpinned the development of policy briefs that took a cultural approach to improving efficiency, warmth and comfort for households, including those in fuel poverty; improving energy efficiency and eco-innovation in businesses; improving uptake of electric vehicles; improving driving efficiency; improving efficiency and emissions from urban freight; increasing uptake of multi-modal mobility; and implementing interventions for a sustainable transport future. Despite the variety of topics, the research revealed generic cultural dynamics that have been discussed earlier, including actors becoming locked into patterns of behaviour and the interplay between internal (actor-initiated) changes and external (often policy-related) influences. Policy settings that consistently support sustainability-oriented motivators, activities and materiality can lead to ongoing journeys oftransformative change, while the wrong settings can direct this journey in an unsustainable direction. Misaligned policy settings are also problematic, as they can set up an environment where there is ambivalence and indeterminacy for sector actors.

Another generic finding across the programme of research was that unsustainable cultures operate at multiple scales, and that the most intractable and problematic cultures may be operating at very broad and influential scales, including within the policy sector. In the Energy Cultures Policy Briefs, we proposed that a culture change was required within New Zealand's various transport policy and implementation agencies in order to achieve a sustainable transport future. To achieve transformational change, those responsible for policy and governance needed to turn their minds to interrogating and adjusting their own cultures—to becoming aware of the ideologies that underpin their work, the languages they use unconsciously, the values implied by their funding decisions and the practices that they have absorbed unquestioningly from their peers.

A Guide to Using the Cultures Framework for Policy Design

Policy development through the lens of the cultures framework invites questions such as: *What features make up the cultural ensembles in the sector we wish to influence? Can we identify different cultural clusters that might require distinctive interventions to avoid inequitable outcomes? Is the outcome desired by policymakers a good 'fit' with the culture/s? Is the culture changing already to give the desired outcomes, and if so what can we do to support that change? If cultural features are poorly aligned with the desired outcomes, might there be a backlash against the policy or unintended consequences? What might be the knock-on effect of changing one aspect of the culture – would other aspects change too, and would that have beneficial outcomes?*

So how might policymakers use cultural analysis to support the development of policy interventions for a sustainable future? In this section, I offer a guide on using the cultures framework for policy development, drawing from my own experiences and those of teams I have worked with, as well as from insights from others' use of the framework as discussed in the first part of this chapter. I begin with a high-level discussion of how the framework can structure a logical process of analysis, followed by a step-by-step guide. I have tried to make this relatively self-explanatory but, for deeper explanations of terms and concepts, readers should refer to Chapter 4.

General Concepts

A policy intervention is often thought of as an intention to change behaviour of sectoral actors to achieve certain outcomes for the greater good. The cultures framework recasts this as an intention to change aspects of an unsustainable culture, which will result in consequential changes to sustainability outcomes. It depicts a policy intervention as a purposeful change to an external influence. To be effective, this needs to result in a change to the relevant actors' cultural ensemble (motivators, activities and/or materiality), and thereby the desired change to the sustainability outcomes.

A premise of the framework is that, in relation to any given outcome, there will be clusters of actors with similar cultural ensembles and similar sustainability outcomes. Previously discussed examples include distinctive cultural clusters in relation to mobility, household energy efficiency and water consumption. Policy analysis using the cultures framework takes an interest in the features of cultural ensembles that have some causal relationship with the issue of concern, any notable heterogeneity across the ensembles and the membership of clusters of similar cultural ensembles.

In investigating culture for policy, we are thus looking not at what is typically counted as culture by lay people, but at patterns of associated motivators, activities and material items that give rise to the outcomes that are of interest. In this sense, the cultural ensembles to be investigated will differ depending on the outcome under study. Cultural characteristics relevant to travel-related greenhouse gas emissions, for example, will likely differ markedly from characteristics relevant to the sustainability of people's food choices. So what culture is for one question is different to what it is for another, and it is important to keep an open mind on this. Rather than looking in the first instance to design policy for a seemingly obvious group of actors, a better starting point would be the outcomes of interest, followed by an exploration of the extent to which different outcomes arise from distinctive sets of cultural characteristics shared by definable groups of actors.

The cultures framework also differentiates the impact of interventions with the terms 'proximal' and 'distal'. These are indicated in Fig. 7.1, where the triangles indicate a change in the feature. Proximal effects refer to changes to the cultural ensemble (motivators, materiality, activities) and/or the extent of the actors' agency. In some instances, interventions that enhance agency may be all that is required to achieve a change in cultural elements and thereby the sustainability outcomes. Distal effects refer to changes in relevant measures of sustainability (e.g. health, equity, economy, environment). By achieving proximal changes, an intervention should achieve distal changes in the sustainability measure/s of interest—if it does not, then it is a failure of policy design. Importantly, changes to the proximal and thereby the distal characteristics must both have positive implications for sustainability measures—there is no point, for example, in interventions that reduce energy consumption (distal effect) while worsening the health and wellbeing of actors because they have turned down their heating (proximal effect). For this reason, policy evaluations should assess both proximal and distal impacts of an intervention.

Although policy work with the framework will generally be seeking to identify cultural ensembles that have a relatively direct relationship with the sustainability outcomes of interest (e.g. energy consumption), there will be other less direct cultural influences on these outcomes (e.g. relating to transport patterns, food expectations or carryovers between work and home). So when we are looking to assess the sustainability implications of particular cultural ensembles, we need to be aware of other cultural features in actors' lives that may work against the intent of policy (for example, parents who are motivated to drive their children to school because of the perceived safety benefits).

When we look at culture from a sustainability perspective, there is no expectation that there will be an idealised 'sustainable culture' towards which all actors will transition. There are many ways in which sustainability outcomes can be achieved, just as there are many different ensembles that deliver unsustainable outcomes. Sometimes cultural ensembles that rank well on one sustainability measure (e.g. low-carbon emissions because of limited energy use) will rank poorly on another (e.g. health and equity measures). Consciousness about sustainability does not necessarily equate to positive measures of sustainability. An environmentally aware wealthy household that adopts symbols of sustainability yet flies overseas for their holidays is likely to have a far higher carbon footprint than a

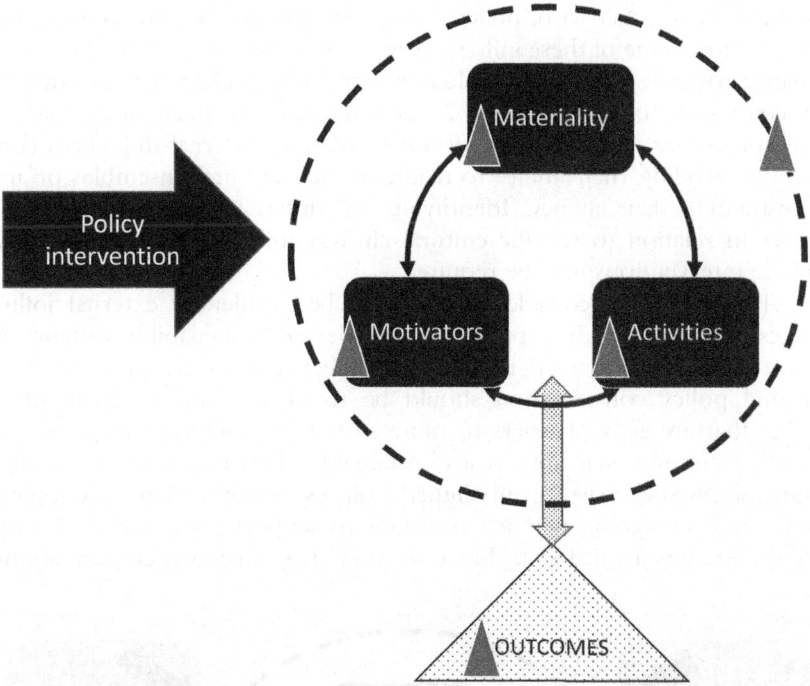

Fig. 7.1 Designing policy with the cultures framework. Any intervention will affect an actor's cultural ensemble and/or their agency (proximal impacts), and this will have consequential implications for measures of sustainability outcomes (distal impacts)

low-income household that cannot afford luxury items or air travel. Analysis through the cultures framework invites a non-judgemental approach to culture—the focus on outcomes means accepting that these could be achieved through any number of diverse cultural arrangements.

Policy as a Change in External Influences

The framework positions policy interventions as changes to external influences on a culture group. As discussed in Chapter 4, external influences may include aspects of the national context and environment, existing policies and laws, institutions, infrastructures and widely shared beliefs and ideologies. As part of policy design, the framework invites consideration of how some of these influences may already be supportive of cultural change towards more sustainable outcomes, while others may be barriers or constraints to change (Fig. 7.2). For cultures that already have sustainable outcomes, it may be important to identify external influences that may be eroding their ability to maintain their cultural ensembles or are constraining their agency. Identifying and differentiating external influences in relation to specific cultural clusters makes it easier to identify where interventions may be required.

The first policy consideration should be to identify external influences that are already supporting the relevant sustainability outcomes: these should be retained unless more effective ones are planned. The second policy consideration should be to identify any external influences that are active barriers to more sustainable cultural change or are eroding cultures that are already sustainable. This may involve dismantling or revising existing misaligned policies as opposed to developing new ones. Designing new interventions to support cultural change may be required as a third step, but only after these first two considerations

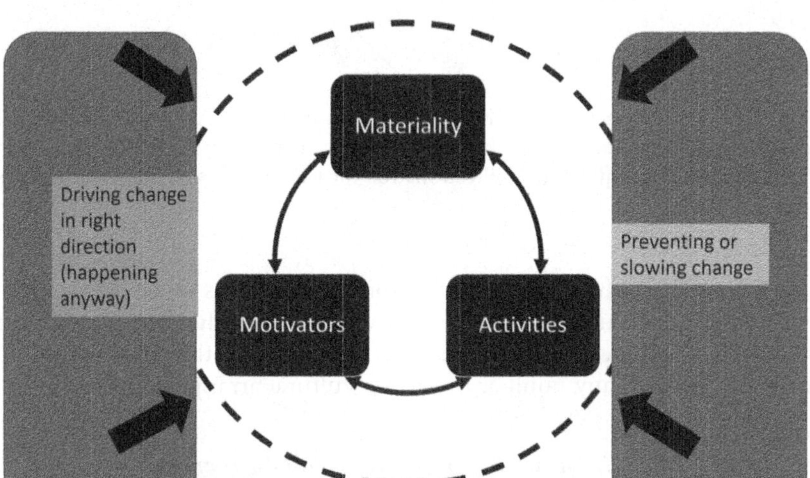

Fig. 7.2 Identifying external influences on a culture. Some influences may already be enabling cultural change in a more sustainable direction, while other influences may be simultaneously preventing or slowing cultural change

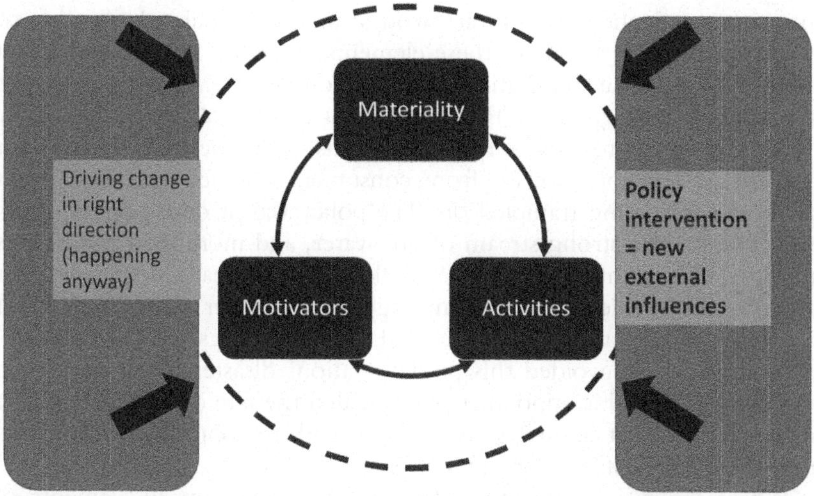

Fig. 7.3 Policy interventions as new external influences on culture

(Fig. 7.3). Together, these initiatives should be designed to have beneficial outcomes for the culture group itself (proximal benefits) as well as for wider sustainability outcomes (distal benefits) (Fig. 7.1).

Changing Cultural Ensembles

Policy does not always need to actively seek to change culture. If there are clear trends whereby groups are already working to become more sustainable, the best policy action may be to remove any barriers and let them get on with it. If groups aspire to become more sustainable but are held back due to agency constraints, then the best policy action may be to simply enhance their agency (e.g. through resourcing or skills development). Active interventions to change cultural ensembles should be a consideration only after these matters have been given careful consideration, which a prior cultural analysis will help with.

Looking at standard policy interventions through a cultures lens, we can see that they generally seek to alter either materiality (e.g. subsidies for electric vehicles), activities (e.g. requiring all businesses to report their greenhouse gas emissions) or motivators (e.g. social marketing campaigns). By showing how motivators, materiality and activities are interconnected, the cultures framework draws policymakers' attention to how a change in any one of these elements can affect other elements. If these consequential implications are not considered, a policy may have unintended consequences. For example, a 2008 policy to only allow 'low-flow' (i.e. water-efficient) shower heads in the New Zealand market resulted in a major backlash from consumers who felt that their right to choose was being trampled on. The policy was at odds with people's expectations of a strong stream of hot water, and more than this, it was a failure of communication, as 'low-flow' showerheads do not in fact produce the trickle that the term suggests. A better understanding of household expectations, languages and understandings relating to showering would have avoided this public relations disaster. As it was, some have argued that this unpopular policy sealed the fate of the government at the election that occurred shortly after, with the opposition using it as a prime example of the 'Nanny State'.

Where policy is at its most effective, it touches on all elements of culture. A good example in New Zealand (although not formally explored through the cultures framework) was a health-related policy programme to reduce the harm of smoking cigarettes, both to smokers and to passive breathers of their smoke. The comprehensive programme introduced policies that changed smoking practices by outlawing smoking from public venues, schools, public transport and businesses open to the public, and other policies that altered material aspects of smoking culture such as outlawing advertising and requiring health warnings and graphic images on cigarette packets. The changes to smoking practices and materiality had consequential impacts on norms, beliefs and values among the general public. This multi-pronged approach has not only markedly reduced the number of smokers in New Zealand but has also led to a massive shift in public norms around smoking, such that today most people would be horrified to see a person smoking in a bar or restaurant.

Working with Cultural Vectors

Policy also needs to pay attention to how culture is learned and shared—the role of cultural vectors. As discussed in Chapter 4, culture is learned through both semantic knowledge and bodily knowledge. Sematic knowledge (know-that) is mainly learned through observational, spoken and written sources, and consists of general understandings of the world. People consciously use these ideas to judge things proper or improper.

Semantic knowledge may be used when people are considering different courses of action, and it is accessed relatively slowly and deliberately to support reasoning, evaluation, judgement and categorisation. Bodily knowledge (know-how), in contrast, is learned through the repetition of bodily actions, repetitive use of perceptual and motor skills, and recurring cognitive and emotional messages. Bodily knowledge is internalised without explicit mediation through thought processes. It is held in the body-mind as a network of associations and accessed via fast pathways that are non-reflective and independent of intention (Lizardo, 2017).

This has clear implications for policy design. Bodily knowledge is usually deployed as a rapid response to a situation and requires little effort, whereas applying semantic knowledge requires a high degree of attention and motivation. People apply bodily knowledge to situations similar to the one in which the relevant associations were formed, while semantic knowledge is able to be applied in a reasoned way to novel situations. Sematic knowledge does not necessarily align with bodily knowledge, and even if sematic knowledge changes, the more automatic bodily responses will remain. This is another reason why 'information deficit' theories of behaviour change often fall short.

Importantly, these different forms of cultural knowledge are not necessarily associated in cognitive processes. One form can be retrieved and used without activating the other, and they are weakly coupled at best. The capacity to make sematic statements does not imply that people have the bodily knowledge to actually enact this perspective, as we saw earlier in the chapter with the example of efficient driving. To be fully effective, policy should be designed in the light of both pathways by which cultural knowledge is learned and retrieved.

Policy and Multi-level Cultures

Culture operates at multiple levels. As discussed in previous chapters, powerful and broadly influential cultures can act as an 'external influence' on cultures that have less agency. When considering policy interventions, it is therefore important to think not only about the culture group that that is most directly linked to the outcome that one seeks to change, but also about more powerful cultures that are shaping or constraining them. For example, to get more citizens using active transport, rather than just seeking to change individual behaviours it might be equally or more important to look at changing the culture of authorities responsible for infrastructure, policy and funding, as their own cultural ensembles will shape their decisions.

A multi-level policy focus was proposed in a paper on achieving sustainable transport in New Zealand (Stephenson et al., 2018). As well as introducing first-level interventions that directly affect transport users, it recommends second-level interventions which involve changes to the processes of decision-making. These drive long-term investments and affect multiple areas of transport such as funding sources, funding allocations and urban form. More deeply, the paper argues for third-level interventions to achieve an all-of-government normative shift so that sustainability principles underpin all transport law and policy, and are also adopted by non-transport agencies (e.g. energy, housing, urban form) whose decisions have implications for future mobility. Once multi-level cultures are aligned, decisions and actions support each other at every level and transformative change is within reach.

Policy Cultures

Policymakers themselves are also embedded in their own culture, and it can be as invisible to them as it is to any other group of actors. Sustainability transitions require a deep rethink of many cultural assumptions, not the least those that are buried in policy culture such as the priority given to technological solutions, the belief that policy should be technology neutral and the dominance of economic theories in policy thinking. Given the urgency and magnitude of the sustainability crisis and the relative ineffectiveness of many policies, it would be helpful for you as a policymaker to investigate the role of your own culture—the implications of your and your colleagues' own biases and practices, and your tendency to replicate past policy approaches even when they don't achieve the desired outcomes. It may be helpful to apply the cultures framework to your policy agency: to identify, for example, shared ways of thinking, dominant forms of knowledge, the use of jargon phrases and traditionally accepted practices. You might also identify the many external influences on your policy culture and explore your capacity to alter it. Ultimately, you should become aware of the extent to which your policy culture itself is a barrier to sustainability transitions.

A Step-by-Step Guide

Having established some high-level ideas about using the cultures framework for policy design, I now use a hypothetical example and take the reader through a step-by-step process of analysis. I focus on the problem of cooling: how to keep households cool as the climate heats up.

1. Clarify Your Intended Outcomes

As a first step, identify the outcomes you are interested in for policy purposes—the triangle at the bottom of Fig. 4.8. For the purposes of our example, the policy problem is that, all else being equal, homes will generally become hotter indoors due to climate change. Households that cannot adjust to this may suffer decreased health and wellbeing, and these consequences are likely to be spread inequitably across the population. Households that have sufficient agency may be in a position to keep their homes cool, thus avoiding health and wellbeing impacts. However, a widespread uptake of air conditioning could lead to a massive increase in peak demand on electricity systems over summer.

Distal outcomes of households' response to a hotter climate might therefore include increased electricity consumption, higher peaks in electricity demand in summer, more travel to cooler climates, or more hospitalisations or deaths from overheating. Proximal outcomes (changes to cultural features) might include householders acquiring more cooling devices, making physical alterations to their homes, changing cooking routines to cooler periods and learning new cooling skills.

From a sustainability perspective, poor outcomes that policymakers may wish to pre-empt range from the implications for individuals (e.g. health impacts) and households (e.g. financial and equity implications) to the impacts on local energy systems (e.g. insufficient supply available from the electricity grid on hot days) and implications for exacerbating climate change (e.g. from increased emissions from the additional electricity generation). An intervention to target just one of these problems won't necessarily solve the others and could well exacerbate them. An integrated policy approach would seek to achieve positive gains on all of these measures as a result of policy interventions.

2. Explore the Current Culture

Having established the outcomes we are interested in, we now seek to understand the existing cooling cultures across the population. Households will already have their own ensembles of material items, activities and motivators that relate to keeping cool on hot days. Relevant materiality might include the way their home is constructed (its insulative properties, windows, etc.), any cooling devices (fans, air conditioners) and the clothing that people wear. Relevant practices may include when and how they use their cooling devices, draw shades against sun intrusion, open windows, drink fluids or take a rest during hot periods. Relevant motivators could include the cooling traditions that come from their upbringing, the knowledge that they have about how to keep cool and their norms regarding a comfortable temperature—for example, 25 degrees may seem cool for some and hot for others.

Households will likely use diverse ways of keeping cool, so to develop policy we first need empirical evidence of these cooling cultures, and whether (and how) they are already changing. One way of eliciting this is through national household surveys. In our research in New Zealand, for example, national surveys enabled us to use cluster analysis to identify four main groups of households with relatively distinctive cultural features in relation to energy efficiency. This could be followed up by in-depth interviews with cluster members to provide more detail.

Even this step might give cause for policy learning. Looking at the different cooling cultures, are there things that policy can learn from households about techniques to keep cool? Are there cooling techniques that households can learn from each other?

If well designed, step 2 should enable policymakers to better understand the heterogeneity of cooling cultures across a population, the different ways in which households are already responding to hotter weather and the variability in outcomes (e.g. how negative outcomes might be unequally spread across a population).

3. Understand the Barriers to Change

Some households will find it easy to adjust their cooling culture. Maybe they can escape the city in summer or can afford more costly energy bills from the use of air conditioning, or can easily take advantage of policies set up to improve insulation and install cooling devices. These households have more agency, and external influences are less constraining on their change. Negative outcomes from these energy culture changes may be more distanced from these households, such as spikes in demand on the electricity grid and increased carbon emissions.

For other households, the capacity to adjust may be highly constrained. This may be because their motivators, activities and materiality relating to cooling are tightly interconnected and thus hard to change. Perhaps they don't have a good understanding of cooling techniques, or maybe other aspects of their everyday life constrain any changes (e.g. gendered roles, cooking routines). External influences beyond their control may constrain their ability to act. For example, if the air outside is polluted or they feel insecure, they might not wish to open windows to create an air flow. It may be that they lack agency—perhaps the capital and running costs of air conditioners is too high, or they can't take advantage of cooling subsidies because they don't own their home. These households may well have aspirations to change but lack the capacity to do so.

Understanding variability in agency is a critical step. Any policy intervention should be designed with an understanding of actors' capacity to alter their cultural ensembles. This includes any constraints on households being able to take advantage of the policy itself, such as language or access barriers. Engaging with householders is the best way to understand constraints on their agency.

4. Consider Multi-level Cultures

An important step in policy development is to identify the various actors or groups of stakeholders that play a role in the outcomes of interest. So far I have only discussed households, but there may well be other actors who influence householder cooling cultures.

As discussed earlier, culture can work on multiple levels to constrain or enable change. For example, living in rental accommodation can strongly constrain households' ability to act. Landlords control much of the material lives of their tenants, and a landlord's motivators, activities and materiality relating to the rental unit will influence their tenants. Research that contrasted landlord and tenant heating cultures in New Zealand revealed that tenants did the best they could but could only make small material changes, and so they had to adjust their expectations of warmth and their heating practices to what was possible within the constraints created by the landlord's heating culture (Nicholas, 2021). These kinds of dynamics are likely to play out in cooling just as much as in heating.

Influential organisations such as councils and government agencies will also constrain or shape the ways households respond to a warmer climate. Their own beliefs, knowledge, languages, activities and material assets can be viewed as another culture that shapes the possibilities for household cooling cultures. Habituation can be as problematic within institutions as it is within households, and possibly more so if this leads to inequitable policy interventions. From a policy perspective, it is important to consider at what level of culture any interventions need to occur.

5. Identify External Influences

This step invites policymakers to identify broader influences that are shaping or constraining change in the cooling cultures. It invites questions such as: What external influences are preventing or slowing cultural change in relation to the outcomes sought? Are some external influences already driving change to this culture?

Mapping external influences (which are likely to vary according to different cooling cultures) helps to target interventions. This may involve actively retaining influences that are already driving positive change and seeking to adjust influences that are limiting household agency and/or are barriers to cultural change. This exercise may involve a regulatory, institutional and policy review, including looking at policies that appear unrelated but still influence the relevant cultures and agency. Engaging with households is also important in order to understand how they are differently affected by external influences.

6. Select the Focus of Your Policy

By now, policymakers should have a good understanding of the heterogeneity in cooling cultures, current cultural changes, unmet aspirations and agency, as well as identification of multi-level cultures and other external influences.

The question now is where (and whether) to intervene. This depends of course on the outcome sought. If the policy intent is to limit increases in electricity demand, then the focus may be households that can afford to acquire and use air conditioning. If the policy focus is on health and well-being, then the focus should be on households that have limited agency to adjust to a hotter climate. A one-size-fits-all policy is unlikely to be effective across all cooling cultures.

Alongside this, it is important not to get caught up in a deficit model of policy development. Are there cooling cultures that should be celebrated and exemplified because they already have positive sustainability outcomes? What can policymakers and others learn from these great examples?

7. Establish Which Aspects of Culture to Influence

Having identified cooling cultures that may need support to achieve policy aspirations, what aspects of that culture and/or members' agency can policy help with? If focusing on householders, don't immediately assume that material changes have to be the place to start. Consider encouraging changes in practices that cost little but make an appreciable difference. Or changes in people's expectations about indoor temperatures. Or maybe help with new understandings and bodily techniques for staying cool. Policy here should be co-developed with householders' own observations and aspirations, and seek to enhance their agency.

Problems may also be identified with more powerful cultural actors. For example, it could be that a key agency is slow to bring in new standards that will assist in keeping homes or urban areas cool, due to misaligned motivators or inordinately slow processes. Changing these may ultimately have a far bigger impact on house temperatures than seeking to change household cultures.

8. Decide on Your Interventions

It is now time to look creatively at the options to support positive change in the selected cooling cultures. The aim of the design process is to achieve proximal change to relevant cultural ensembles and thus achieve distal change to the desired sustainability measures. This means staying aware of two moving targets at the same time—considering the broader implications of proximal change (e.g. will the intervention impact actors' health and wellbeing in unintended ways?) as well as for the distal goals (e.g. will it also reduce greenhouse gas emissions?). Interventions to achieve one outcome (e.g. reduce peak demand from air conditioning) could have negative consequences for other outcomes (e.g. health and wellbeing, equity) so multi-focused policy design may be needed.

It should be noted that culture change isn't always a desirable outcome. As already noted, some cultures may already be strongly aligned with the outcomes sought, in which case policies may seek to ensure that those cultural characteristics are supported and reinforced. It is also important to retain and enhance any external influences that are already supporting effective cooling cultures. In other instances, effective household cooling cultures may be latent, in which case the first place to look for interventions is to remove any barriers to positive changes to which householders already aspire, but are held back by their limited agency and/or external influences.

Beyond this, targeted interventions may be required to support the kinds of changes identified in step 7. These fall more in the arena of 'normal' policy design, but even here, awareness of the dynamics of cultural change is critical. A change in one aspect of culture may lead to consequential changes for other aspects of culture, which could have unintended consequences. Some trials that seek to understand these consequential changes would be advisable.

Don't forget also that it may be more important to target cultural change at other levels—within institutions or among landlords for example—rather than assuming that households are the 'problem'. Good policy design forsustainable outcomes needs to recognise and address cultural change at all levels (even within the policy agency itself) in order that all cultural shifts are aligned with the outcomes, rather than working against each other.

9. Evaluate

Good policy design includes post-intervention evaluation of its effectiveness. Evaluation is a critical part of any policy cycle and involves having measures of the situation prior to any interventions, and follow-up measures of the same indicators after the policy has been in place for some time. Evaluation is particularly important from a cultural perspective because the same intervention may have different outcomes depending on the households and their existing cultural ensembles and agency. Even if it has not been used to design interventions, the cultures framework can be used to underpin policy evaluations.

Traditionally, evaluation has tended to focus on 'outcome measures'—for example, has water consumption decreased, is electricity consumption more efficient, are there fewer hospital admissions, are there more fish in the stream? In the language of the cultures framework, these are distal measures. In contrast, the framework invites evaluation of the impact of an intervention on the relevant cultural ensemble (proximal impacts), *as well as* on any measurable shift towards the goal of the intervention (distal impacts) (see Fig. 7.1). For a cooling-related policy intervention with households, for example, measurement of proximal outcomes might include whether and how households have changed their cooling practices, skills, norms and understandings, and/or any physical changes to their home or appliances, and/or changes to their agency. Measurement of distal outcomes might include any change in energy consumption arising from these cultural shifts, or changes to measures of health and wellbeing.

A dual focus on proximal and distal outcomes can help reveal why interventions may not be as successful as hoped. For example, the study in Ireland on energy efficiency interventions in social housing (discussed in Chapter 5) showed that the targeted increases in thermal comfort weren't achieved in some households due to the persistence of old routines, while other households turned up the heating and thus made no savings, thus foiling two of the policy goals of the intervention (Rau et al., 2020). Policy can also fail because it has a poor fit with the culture it is attempting to influence: in Chapter 5, I described the US Navy's attempts to introduce energy-efficient lighting which was foiled by naval norms, materiality and practices that resisted change (Dew et al., 2017). On the other hand, policy can be far more influential than anticipated, as with the uptake of solar lighting in Vanuatu described in Chapter 6, with consequential changes to energy culture and other aspects of daily life that had not been anticipated (Walton et al., 2014). Evaluating both proximal and distal changes can help reveal how and why interventions are successful or not, and whether unintended consequences have resulted, and thereby can help with adjusting interventions to achieve better outcomes.

Some examples of how the cultures framework has been used to evaluate the effect of interventions have already been introduced. These include a comparison of two different kinds of interventions to improve household energy efficiency in New Zealand (Scott et al., 2016), an investigation into the failure of top-down interventions to change cooking methods in Zambia (Jürisoo et al., 2019) and an assessment of the effect on household energy culture of interventions to reduce or shift electricity consumption in Canada (Lazowski et al., 2018).

To date, most evaluations of proximal outcomes have used pre- and post-intervention surveys. However, there have been attempts to standardise measures of cultural change. In the energy field, two consecutive papers developed and tested an evaluation toolkit for household energy interventions that was based on the elements of the cultures framework. The authors' focus was on behaviour-based energy interventions that aimed to reduce energy consumption. This work developed an empirically verified set of measures to evaluate energy culture before and after an intervention. The measurement instruments (questions developed based on behavioural theories) were tested to ensure they were reliably interpreted, measured the constructs they were intended to and predicted behavioural intentions (Ford et al., 2016; Karlin et al., 2015). The toolkit was developed for implementation in California and the survey questions were designed for this context, so its use elsewhere would require some adjustments of the evaluation instruments.

Evaluation of proximal as well as distal outcomes of interventions can thus provide deeper learning as to how interventions affect the features and dynamics of culture, as well as any sustainability outcomes of these changes. This can support programme improvement and enable comparison between different types of interventions.

Interventions for Change by Businesses, Organisations and Communities

It is not only in the policy world that people need to make decisions about how to stimulate change. Many people in businesses, organisations and communities are also motivated to become more sustainable. The cultures framework has been used in several studies of businesses to help them to understand their energy culture, and from there to help them to determine how to tailor interventions to achieve a more sustainable culture.

Interventions for Sustainable Energy Use in Small-Medium Enterprises

A good example is a study on business energy cultures that was part of a European Union Horizons 2020 research project. The project assessed business energy culture and potential interventions for more sustainable outcomes. It was designed to be used by energy managers and others to evaluate the state of energy culture and engage employees in more sustainable energy-related practices. A survey, based on the elements of the cultures framework, was initially tested with expert participants in Finland, Italy, Switzerland, Germany, France and Austria, and subsequently applied in 65 small-medium businesses with more than 20 employees, located in the six partner countries.

The survey contained 13 questions for energy managers and 12 for employees. Survey questions for staff within the business were designed around five topics aligned with the cultures framework: awareness of energy technologies within the firm; personal beliefs, aspirations and motivations relating to sustainability; current energy practices in the company; external factors that shape energy culture at work, including national regulatory frameworks; and participants' perceptions of barriers to change. The survey sought to capture the many complex elements involved in the nature of energy use within a firm, and was designed to help energy managers understand the relative 'maturity' of the company's energy culture and identify where change was needed. It was also seen as a way of enabling all employees to understand the firm's energy culture and their role in it, so that they could be active participants in change. Based on the firms' scores in the 'maturity matrix' relating to their respective energy cultures, they were provided with tailored advice based on expert recommendations developed by the six partner countries (Fatima et al., 2021; Oksman et al., 2021).

Designing a Community Energy Management Programme

The framework can also help to identify opportunities for action or intervention in community-based initiatives. This example is from research with households in an eco-conscious mixed-use urban village in Austin, Texas, which had high levels of solar PV and electric vehicles. One of the problems with the uptake of distributed renewable generation is that it can have repercussions for the electricity grid, one of which is the 'duck curve' that occurs when solar generation ends at sunset and households all start using more power from the grid at the same time, resulting in a steep upward curve that looks a bit like a duck's neck. This study looked at how the duck curve could be flattened if community members worked together to adjust the timing of their energy use for collective benefits. The project's starting point was that any energy management programme, including energy feedback tools, should be designed around the community's social and physical context—its energy culture. The researchers used the cultures framework to reveal motivations for participating in a collaborative energy programme.

Their study found households had widely shared social norms that aligned with collective action to design and implement a local energy management programme, even though their material culture and practices were quite varied. This suggested to the researchers that the programme should emphasise societal goals and environmental impacts, and these were incorporated into a co-design process that involved researchers, designers and potential users. Understanding the energy culture helped with this process by prioritising particular criteria that could sustain and strengthen community engagement. Overall, the model they developed to represent the influence of culture was able to predict about 46% of the variance in favourable attitudes towards the proposed programme (Krietemeyer et al., 2021).

I am also aware of other non-policy contexts in which the cultures framework has been used by community-based organisations in New Zealand and internationally. Some have used it to assist their community or organisation to develop an awareness of their own cultural ensembles, while others have used it to design three-dimensional programmes of action that seek to influence motivators, materiality and activities for more sustainable outcomes. As far as I am aware, none of these have been formally written up, but it is pleasing to hear that this approach is useful. The cultures framework can be easily understood by both technical and non-technical audiences, which is vital for encouraging participation and motivation. I am excited by the potential for the cultures framework to be more widely used in lay contexts.

Conclusion

This chapter has mostly focused on the use of the cultures framework to underpin policy analysis, the design of interventions and evaluation. From the outset, the framework has shown itself to be particularly useful in this respect: supporting the development of comprehensive policy advice in the energy cultures programmes, and of policy recommendations arising from many of the research articles discussed in this book. These experiences have emboldened me to suggest that it could be more widely used as a policy framework.

When seeking to design interventions for better sustainability outcomes, one certainty is that culture matters. Culture offers a deeper understanding of how and why humans or organisations act as they do, and has much to offer for the design and evaluation of policy interventions. Cultural analysis can assist in the identification of patterns of cultural ensembles across a population. It can reveal how cultural attributes can shape different responses to interventions and result in unintended outcomes, and why some policies may have regressive impacts on some culture groups. It can help reveal constraints in actors' agency and draw attention to the multiple influences at play that shape their cultural ensembles and their ability to respond to a policy intervention. It complements policy approaches that see society as comprised of individuals, identifying broader influences on behaviour than economic maximisation and broadly shared psychological traits. In a world where 'information deficit', 'rational actors' and 'nudge theory' dominate policy conversations, culture offers more nuanced understandings and reveals new opportunities for policy action.

The cultures framework offers conceptual contributions to policy development by bringing notions of cultural ensembles, agency, external influences and multi-level cultures into the policy arena, and proposing the evaluation of both proximal and distal outcomes. I have outlined a generic process of policy development that could be tailored to different circumstances. However, I need to emphasise that it is a relatively novel approach and to date, as far as I am aware, has only been used by the research community. It would be good to see this approach applied and evaluated by policy agencies in the future.

References

Barton, B., Blackwell, S., Carrington, G., Ford, R., Lawson, R., Stephenson, J., Thorsnes, P., & Williams, J. (2013). *Energy cultures: Implication for policymakers*. Centre for Sustainability, University of Otago.

Berry, F. S., & Berry, W. D. (2018). Innovation and diffusion models in policy research. In C. M. Weible & P. A. Sabatier (Eds.), *Theories of the policy process* (4th ed., pp. 253–297). Routledge.

Borrás, S., & Edler, J. (2020). The roles of the state in the governance of socio-technical systems' transformation. *Research Policy, 49*(5), 103971.

Carrus, G., Tiberio, L., Panno, A., Lettmayer, G., Kaltenegger, I., Demir, M. H., Biresselioglu, M. E., Solak, B., Dimitrova, E., Tasheva-Petrova, M., & Mutafchiiska, I. (2019). *Analysis of enabling factors for consumer action*. ECHOES Research report, NTNU, Norway.

Cherp, A., Vinichenko, V., Jewell, J., Brutschin, E., & Sovacool, B. (2018). Integrating techno-economic, socio-technical and political perspectives on national energy transitions: A meta-theoretical framework. *Energy Research and Social Science, 37*, 175–190.

Cortés, A., & Amigo, C. (2022). Energy culture and the dynamics of energy poverty in south Chile: A blind spot for decontamination energy efficiency policies. *People, Place and Policy, 16*(1), 73–97.

Debnath, R., Bardhan, R., Darby, S., Mohaddes, K., Sunikka-Blank, M., Coelho, A. C. V., & Isa, A. (2021). Words against injustices: A deep narrative analysis of energy cultures in poverty of Abuja, Mumbai and Rio de Janeiro. *Energy Research & Social Science, 72*, 101892.

Dew, N., Aten, K., & Ferrer, G. (2017). How many admirals does it take to change a light bulb? Organizational innovation, energy efficiency, and the United States Navy's battle over LED lighting. *Energy Research & Social Science, 27*, 57–67.

Fatima, Z., Oksman, V., & Lahdelma, R. (2021). Enabling Small Medium Enterprises (SMEs) to become leaders in energy efficiency using a continuous maturity matrix. *Sustainability, 13*(18), 10108.

Ford, R., Karlin, B., & Frantz, C. (2016). Evaluating energy culture: Identifying and validating measures for behaviour-based energy interventions. In *Proceedings of the International Energy Policies & Programmes Evaluation Conference* (pp. 1–12).

Ford, R., Walton, S., Stephenson, J., Rees, D., Scott, M., King, G., Williams, J., & Wooliscroft, B. (2017). Emerging energy transitions: PV uptake beyond subsidies. *Technological Forecasting and Social Change, 117*, 138–150.

Geels, F. W. (2002). Technological transitions as evolutionary reconfiguration processes: A multi-level perspective and a case-study. *Research Policy, 31*(8–9), 1257–1274.

Geels, F. W. (2005). Processes and patterns in transitions and system innovations: Refining the co-evolutionary multi-level perspective. *Technological Forecasting and Social Change, 72*(6 SPEC. ISS.), 681–696.

Geels, F. W. (2011). The multi-level perspective on sustainability transitions: Responses to seven criticisms. *Environmental Innovation and Societal Transitions, 1*(1), 24–40.

Geels, F. W. (2012). A socio-technical analysis of low-carbon transitions: Introducing the multi-level perspective into transport studies. *Journal of Transport Geography, 24*, 471–482.

Geels, F. W., Pieters, T., & Snelders, S. (2007). Cultural enthusiasm, resistance and the societal embedding of new technologies: Psychotropic drugs in the 20th century. *Technology Analysis and Strategic Management, 19*(2), 145–165.

Geels, F. W., Sovacool, B. K., Schwanen, T., & Sorrell, S. (2017). The Socio-Technical Dynamics of Low-Carbon Transitions. *Joule, 1*, 463–479.

Geels, F. W., & Verhees, B. (2011). Cultural legitimacy and framing struggles in innovation journeys: A cultural-performative perspective and a case study of Dutch nuclear energy (1945–1986). *Technological Forecasting & Social Change, 78*(6), 910–930.

Jürisoo, M., Serenje, N., Mwila, F., Lambe, F., & Osborne, M. (2019). Old habits die hard: Using the energy cultures framework to understand drivers of household-level energy transitions in urban Zambia. *Energy Research and Social Science, 53,* 59–67.

Karlin, B., Ford, R., & McPherson Frantz, C. (2015). Exploring deep savings: A toolkit for assessing behavior-based energy interventions. In *Proceedings of the International Energy Program Evaluation Conference* (pp. 1–12).

Khan, I. (2021). Factors dominating peak electricity demand in Bangladeshi urban households: An assessment through the *energy cultures framework. Energy Sources, Part B: Economics, Planning, and Policy, 16*(3), 279–299.

Krietemeyer, B., Dedrick, J., Sabaghian, E., & Rakha, T. (2021). Managing the duck curve: Energy culture and participation in local energy management programs in the United States. *Energy Research & Social Science, 79,* 102055.

Lazowski, B., Parker, P., & Rowlands, I. H. (2018). Towards a smart and sustainable residential energy culture: Assessing participant feedback from a long-term smart grid pilot project. *Energy, Sustainability and Society, 8*(1), 1–21.

Lizardo, O. (2017). Improving cultural analysis: Considering personal culture in its declarative and nondeclarative modes. *American Sociological Review, 82*(1), 88–115.

Manouseli, D., Anderson, B., & Nagarajan, M. (2018). Domestic water demand during droughts in temperate climates: Synthesising evidence for an integrated framework. *Water Resources Management, 32*(2), 433–447.

McKague, F., Lawson, R., Scott, M., & Wooliscroft, B. (2019). Understanding energy poverty through the energy cultures framework. In Simcock, N. Thomson, H. Petrova, S. Bouzarovski, S. (Eds), *Energy Poverty and Vulnerability: a Global Perspective* (pp. 33–45). Routledge.

Nicholas, L. (2021). *On Common Ground: How landlords & tenants shape thermal performance of private rental properties* [Masters Thesis], University of Otago.

Oksman, V., Reda, F., Karjalainen, S., & Fatima, Z. (2021). Towards sustainable energy culture in the industrial sector: Introducing an interdisciplinary method for understanding energy culture in business industries. *Energy, Sustainability and Society, 11*(1), 1–13.

Rau, H., Moran, P., Manton, R., & Goggins, J. (2020). Changing energy cultures? Household energy use before and after a building energy efficiency retrofit. *Sustainable Cities and Society, 54,* 101983.

Reisch, L. A. (2017). Behavioural economics, consumer behaviour and consumer policy: State of the art. *Behavioural Public Policy, 1*(2), 190–206.

Russell-Bennett, R., McAndrew, R., Gordon, R., & Letheren, K. (2019). Effectiveness of Household Energy Efficiency Interventions in Advanced Economies: What works and what doesn't. Report for Energy Consumers Australia. *Group of Energy Efficiency Researchers, Australia.*

Sahakian, M., & Wilhite, H. (2014). Making practice theory practicable: Towards more sustainable forms of consumption. *Journal of Consumer Culture, 14*(1), 25–44.

Sandfort, J. R., & Moulton, S. (2020). Replication or innovation? Structuration in policy implementation. *Perspectives on Public Management and Governance, 3*(2), 141–154.

Scott, M. G., & Lawson, R. (2018). The road code: Encouraging more efficient driving practices in New Zealand. *Energy Efficiency, 11*(7), 1617–1626.

Scott, M. G., McCarthy, A., Ford, R., Stephenson, J., & Gorrie, S. (2016). Evaluating the impact of energy interventions: home audits vs. community events. *Energy Efficiency, 9*(6), 1221–1240.

Shove, E. (2014). Putting practice into policy: Reconfiguring questions of consumption and climate change. *Contemporary Social Science, 9*(4), 415–429.

Sovacool, B. K., & Griffiths, S. (2020). The cultural barriers to a low-carbon future: A review of six mobility and energy transitions across 28 countries. *Renewable and Sustainable Energy Reviews, 119*, 109569.

Stephenson, J., Barton, B., Carrington, G., Hopkins, D., Lavelle, M. J., Lawson, R., Rees, D., Scott, M., Thorsnes, P., Walton, S., & Wooliscroft, B. (2016). *Energy cultures policy briefs*. Centre for Sustainability, University of Otago.

Stephenson, J., Spector, S., Hopkins, D., & McCarthy, A. (2018). Deep interventions for a sustainable transport future. *Transportation Research Part D: Transport and Environment,* 61, 356–372.

Thaler, R. H., & Sunstein, C. R. (2008). *Nudge: Improving decisions about health, wealth, and happiness*. Yale University Press.

Walton, S., Doering, A., Gabriel, C.-A., & Ford, R. (2014). *Energy transitions: Lighting in vanuatu*. Project report, Australian Aid.

CHAPTER 8

Using the Cultures Framework for Research

Introduction

I cannot think of a single research topic relating to the human/sustainability nexus that does not have a cultural component. From globally influential paradigms and the practices of fossil fuel majors to the operations of small businesses and the daily lives of households, culture is involved. Yet as a research topic, culture is often remaindered—applied as a loose label for a collection of features of social existence that sit unexamined alongside other deeply analysed phenomena. Apart from the fraction of researchers trained in forms of cultural analysis, the slippery, qualitative features of culture can seem too hard to investigate by researchers interested in sustainability issues. The cultures framework addresses this difficulty by offering a structured way to approach cultural research that can be used by researchers from almost any disciplinary background. As described in earlier chapters, the framework has been sufficiently tested to have confidence that it can fruitfully guide research endeavours.

Over the 12 years since the framework was first introduced in the literature, it has been used with a broad range of research approaches. It has been used with qualitative methods, quantitative methods and mixed methods. It has been used to formulate research design, as an analytical frame for the interpretation of existing data sets, and as a conceptual framing for meta-reviews of data. It has been used by individual researchers from a single discipline as well as by interdisciplinary

J. Stephenson, *Culture and Sustainability*,
https://doi.org/10.1007/978-3-031-25515-1_8

research teams. It has underpinned undergraduate studies, postgraduate dissertations and extensive research programmes. It has been used to design research-based interventions and as an evaluation framework. And as shown in earlier chapters, it has been applied to a wide variety of problems and fields of enquiry.

This chapter is designed to guide academics and students who wish to undertake research using the cultures framework. For most of the chapter, I discuss the use of the framework to explore the interplay between culture and sustainability. By providing a structure for research investigations, the framework can help reveal what cultural ensembles, consisting of what cultural features, have causal relationships with what outcomes, affected by what external influences. It can help determine who are the actors within the culture group under study, and which cultural ensembles are already more sustainable than others. By examining the internal dynamics of culture, we can see how this leads to the sustainability outcomes. By studying cultural ensembles in relation to external influences, we can gain insights into why cultures remain static or evolve. By investigating the scope of agency of cultural actors, we can better understand why it is difficult for them to change, and who comprise more powerful organisations or institutions. And by examining whether a culture is dynamically stable or has the potential to change, we can gain insights into whether transformation is possible. Of course, not all these questions will be relevant to a given study, and the choice of questions will be determined by the particular context of the research and the sustainability issues at stake, but this gives an indication of the types of questions for which the framework can be used.

Towards the end of the chapter, I discuss how the framework can also be used as a meta-theoretical framing. In this sense, it can be used as an overarching structuring device for multidisciplinary, multi-theoretical and multi-method research, as with the examples of the Energy Cultures research programmes discussed in Chapter 7. As covered in Chapter 4, it builds on mature social science and cultural theories, and the framework acts as a structuring device for reaching into these fields of knowledge to examine dynamics and causal mechanisms in greater depth. I finish by discussing how the diverse and currently fragmented cultural theories discussed in Chapter 2 can make a stronger contribution to sustainability research.

Core Concepts

On the assumption that some readers may skip directly to this chapter, I will first recap on some key concepts. First, what culture is not. Culture is not about how people operate as individuals, each with their unique personal history and psychology. It is not about demographics. It is not about features that all humans share as social beings. All of these may interplay with culture, but they are not its defining characteristics. Second, the cultures framework is not just about culture. Cultures do not exist in a vacuum, and the framework draws attention to important variables that shape culture and mediate its implications for sustainability outcomes.

Recapping on what I mean by culture in this context, I describe it in Chapter 4 as comprising *distinctive patterns of motivators (norms, values, beliefs, knowledge and symbolism), activities (routines and actions) and materiality (products and acquisitions) that form dynamic ensembles which are shared by a group of people and learned through both cognitive and bodily processes*. These cultural ensembles can be most simply described as similar ways of thinking, doing and having that are evident across a group of people. Depending on the focus of research, this could apply to the cultures of people in their everyday lives, or the cultures of organisations or businesses, or cultures at even broader scales of institutions and ideologies.

Rather than focusing on describing groups that we typically think of in cultural terms (e.g. ethnic cultures, youth cultures, American culture), the framework invites inquiries into actors and their cultural ensembles that have implications for sustainability. Relevant actors may be identified at any scale: individuals, households, communities, organisations and beyond. Culture can be investigated in relation to a sustainability problem in both a causal sense and in the way in which cultural dynamics can resist change. Culture can also be investigated as part of sustainability solutions in the sense that many existing cultures are exemplars of sustainability. Cultural change can be a creative and fast-moving force for sustainability transitions.

The range of concerns of sustainability is exemplified by the United Nations' 17 Sustainable Development Goals (United Nations, 2015) but as discussed in Chapter 1, the SDGs are only one perspective on sustainability. More critical perspectives suggest that sustainability will not be achieved without more radical change to established systems of production and expectations of consumption. The framework does not predetermine what is meant by sustainability outcomes—this is left to the researcher to determine for their particular context.

The cultures framework has evolved over time. All the research examples I included in previous chapters used earlier terminologies ('energy cultures framework' or 'sustainability cultures framework') but the core concepts have changed little over time apart from becoming more generic. Earlier versions have produced sound and fruitful findings, and there is no reason why researchers cannot continue to use it in its earlier and slightly simpler form, especially those wishing to take their first forays into this field, in which case the guidance in Stephenson (2018, 2020) will be helpful. In this chapter, I continue to use examples from this prior research, but describe how researchers can undertake inquiries using the revised cultures framework which is presented in Chapter 4. I encourage readers of this chapter to return to Chapter 4 for fuller descriptions of the language, elements and dynamics of the revised framework.

A Guide to Research with the Cultures Framework

The framework can be applied in many different ways to support research processes. One way is to simply use the diagram of the cultural ensemble (Fig. 8.2) as the basis for describing a culture—its distinctive elements and their dynamics. It can be surprisingly difficult to explain what culture is, and these concepts give a solid foundation for identifying cultural features in any given context. At my university, some lecturers ask students to undertake at-home research to describe their own cultural ensembles that relate to energy use or greenhouse gas emissions. The diagram showing the ensemble, agency barrier and external influences (Fig. 8.5) has been used as the basis for discussions about research culture in university departments, and by research organisations to analyse how their own culture may be holding them back from undertaking transformative research. In these instances, it can be a tool for self-reflection, enabling actors to understand and articulate elements of their culture and constraints on change.

But more commonly, the framework is used by researchers in its full form (Fig. 8.1) for sustainability-related investigations. Some studies primarily seek to describe the cultural characteristics of a particular population. Examples I have discussed in earlier chapters include cooking cultures in Zambia, energy cultures in rural households in Transylvania and mobility cultures in New Zealand. Studies have also used the framework to compare cultures within a population. These have generally sought to explore the contribution of cultural differences to sustainability outcomes. Examples discussed in previous chapters include identifying varied cultural ensembles across populations in relation to energy consumption, energy efficiency and water consumption. Others have explored aspects of culture as influences on people's readiness to engage in new collective behaviours, their responses to efficiency retrofits and as a factor in nations' willingness to decarbonise.

The framework is also useful for exploring barriers to cultural change. Studies discussed in earlier chapters have identified cultural characteristics that help explain resistance to change in the US Navy, failures to achieve desired levels of change in social housing interventions and cultural barriers to change in academic air travel. Chapter 7 discusses at length how the framework can be used as a basis for policy development and to underpin the evaluation of interventions.

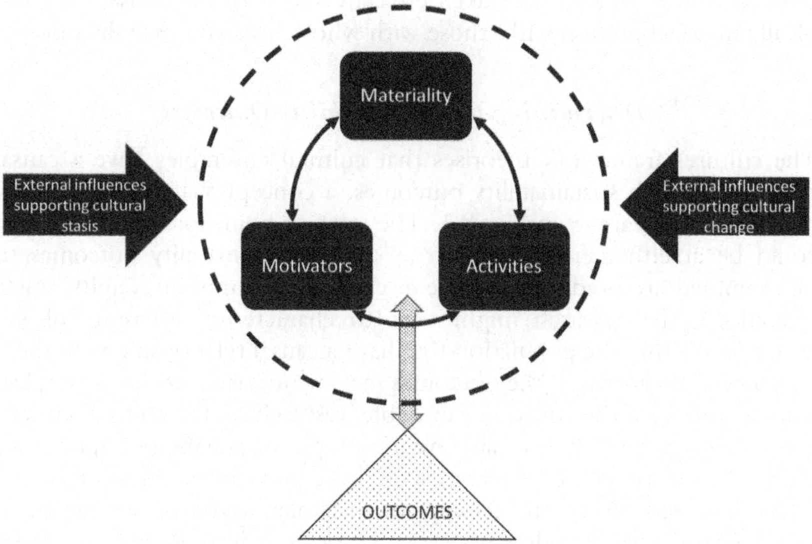

Fig. 8.1 The cultures framework

The following section describes how to use the revised cultures framework to underpin research. For easier reference, I repeat here (as Fig. 8.1) the complete cultures framework diagram (first appearing as Fig. 4.8 in Chapter 4). The section is ordered as a step-by-step process, although it should be noted that not all research will involve all stages, and some research may proceed in a different order or head in different directions. Following this, I describe the range of research methods that have so far been used with the framework, and its methodological inclusivity in general.

Research on culture is research with people. In exploring what is needed to achieve societal transitions towards sustainability, researchers might wish to learn from groups and organisations that have already grappled with what it takes to live sustainably. They may wish to explore unsustainable cultures that seem unlikely to change. They might seek to work with culture groups or organisations that wish to change but can't, or with those that are already on change journeys. In any situation, the research process and its outcomes have the potential to destabilise established beliefs, ways of life and social processes. Social research is a serious business and must be undertaken ethically and with the consent of, and ideally in collaboration with, those with whose lives you may disrupt.

Determining the Sustainability Outcomes

The cultures framework theorises that cultural ensembles have a causal relationship with sustainability outcomes, a concept that is conveyed by the two-headed arrow in Fig. 8.1. The starting point for research design could be at either end of the arrow. If the sustainability outcomes to be examined are predetermined (e.g. energy consumption, equity, waste reduction), the research might seek to characterise different cultural ensembles within the population that have a causal relationship with these outcomes. Alternatively, the outcomes may be uncertain at the outset, but will emerge from the study. For example, research on the cultural ensembles of elderly households may reveal multiple sustainability implications such as health outcomes, energy expenditure and carbon emissions.

As discussed earlier, the concept of sustainability outcomes can be as broad or as specific, and as conservative or as radical, as the researcher wishes to make it. To be useful for the purposes of the framework, outcomes ideally are measurable (i.e. empirical evidence is available as to whether that outcome is improving or degrading) or at least able to be qualitatively described and compared. Outcomes can be uni-dimensional (e.g. a measure of water quality) or might consist of multiple interconnected qualities (e.g. health, biodiversity, equity). Outcomes may be of widespread benefit (e.g. reducing greenhouse gas emissions) or directly beneficial to the households themselves (e.g. improved health).

The double-headed arrow between cultural ensembles and outcomes also reminds researchers that if outcomes change, this changed context can become a further external influence. For example, if a farmer introduces practices that result in cleaner rivers and streams, this may create positive reinforcement for further cultural change. The farmer may enjoy being able to catch fish again, or seeing their children swimming, or hear positive feedback from community members, and may be encouraged to do more. Research on this kind of feedback could help identify whether and how positive affirmation from more sustainable outcomes can lead to ongoing cultural transformation.

Determining the Cultural Elements and Their Interactions

At an early point in the research process, it will be necessary to determine both the scope of the cultural ensemble and the scope of the member actors. The cultural elements to be studied will ultimately be determined by the sustainability outcomes you are interested in and the actors you are focusing on. For example, if you are interested in carbon emissions from a business sector, the obvious cultural actors to focus on would be those businesses, and the elements to study would be the motivators, materiality and activities that have a direct relationship to carbon emissions. However, from identifying this first-order group of actors, it may become clear that other actors also play a role. It may prove more useful to focus on a sub-group within the business such as senior leadership, or shareholders, or alternatively it may prove important to examine cultural factors at broader scales, such as at the sector level, or within suppliers for these businesses. As a researcher, be open to which group/s of actors it might be most useful to focus on. Depending on the research aim, it might be more useful to gain a rich understanding of the cultural ensembles of a small number of actors, or alternatively to investigate a narrow range of cultural features across a much larger population.

You may find it is useful to examine cultures at multiple scales. Cultures are identifiable and discussable from a minute scale, such as the cultural ensemble of a particular actor or organisation, to massive scales, such as the distinctive and enduring features of Western civilisation. You may find it useful to study the ways in which culture can act as structure—a high-level ensemble of beliefs, symbolism, practices and institutions, which shape other cultures that have less power and reach. The framework is relevant to supporting research at any scale and scope.

The next step is to determine which cultural features are most relevant to your study. The core variables of the framework reflect concepts about culture that repeatedly appear in cultural theories (see Chapter 3). *Materiality* comprises items that are made, acquired, owned, accumulated, held or nurtured by cultural actors. *Activities* are frequent and infrequent actions undertaken by cultural actors. *Motivators* are shared aspects of cognition that include norms, values, beliefs, knowledge and meanings (Fig. 8.2). Which specific features comprise the ensemble for the purposes of your research will depend on the sustainability outcomes, the nature of the actors and of course your interests as a researcher.

I have described a cultural ensemble as a generally consistent pattern of materiality, motivators and activities displayed by an actor or group of actors, but all three may not have equal pertinence depending on the issue—for example, beliefs, meanings or understandings may be more relevant in a particular case than activities or materiality, and vice versa. Cultures will rarely be distinguishable by unique sets of cultural elements; there may be a great deal of overlap between the ensembles of culture groups. The ways in which they are differentiated will be determined by the research context.

When doing research with the cultures framework, how deeply to drill into each element will depend on the nature of the study. For example, in relation to 'motivators', many studies to date have focused on depicting norms, because this was the terminology of the original cultures framework. Other studies using the framework have explored morals, meanings, values, knowledge and beliefs, which is one of the reasons for replacing 'norms' with 'motivators' in the revised cultures framework. In terms of activities, some studies have focused on routines while others have been more interested in one-off or occasional actions. Some have centred on one type of material item, while others have been interested in material assemblages. The nature of the topic should shape the focus of the research, and the researcher should hold open the possibility that relevant but unsuspected cultural features may emerge in the course of the research.

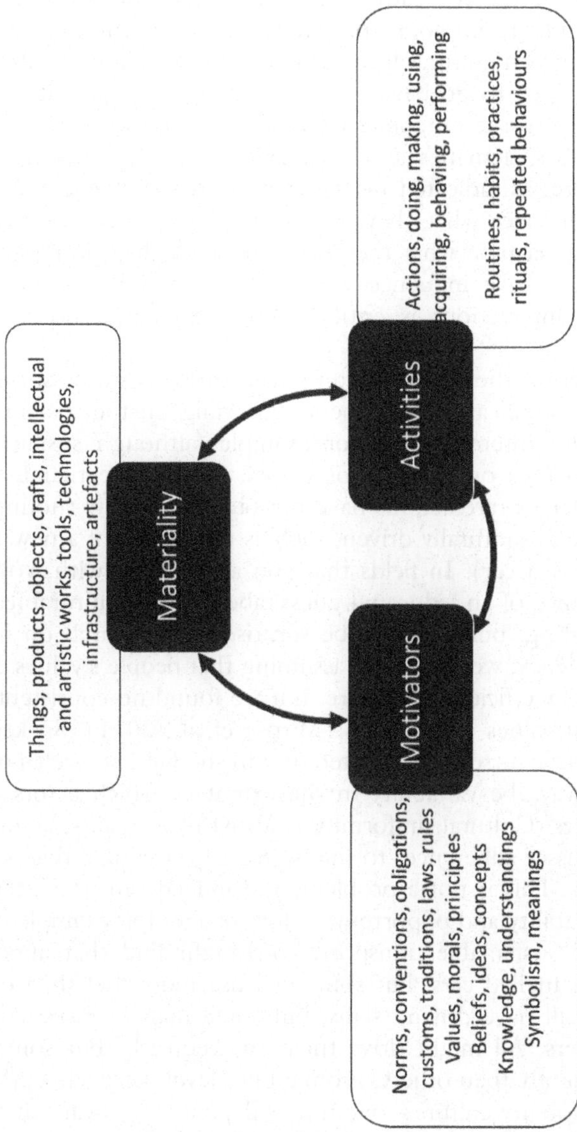

Fig. 8.2 The cultural ensemble—the core elements and their dynamics

Although the research process will likely identify specific cultural features that can be directly causally related to the outcomes (e.g. the presence of particular technologies or practices), it is the cultural ensemble and dynamics between cultural features that make it 'cultural'. Research should therefore seek to go beyond simply listing cultural features that bear a relationship to the outcomes of interest, to considering how they interact. Motivators, activities and materiality form the interconnected 'system' of culture, as indicated by the curved arrows in Fig. 8.2. How people think influences what they acquire and how they act; people's activities partly determine what they have and how they think; and the things that people have influence what they do and how they think. Exploring these interactions is critical to understanding how culture operates.

When considering the scope of cultural features to study, some may emerge as more significant than others depending on the sustainability outcomes you are interested in. For example, although sustainability research often focuses on routines or habits, it could be that in some instances, irregular or rare actions have the biggest impacts (noting that these can be equally culturally driven, such as the choice of a new house or whether to buy a car). In fields that you are familiar with, you may have a better chance of an 'educated guess' about which cultural features to start investigating, but you may be surprised. In research on household energy efficiency, we started by assuming that people's values would strongly shape how efficient they were, but we found no consistent relationship between values and actions (Mirosa et al., 2011). So keep an open mind and, of course, explore literature in the field beforehand.

There will always be variability in the extent to which actors adopt cultural ensembles. Cultural uniformity is a myth—in reality, actors will have greater or lesser adherences to the 'signature' ensemble that is identified in research. This is not a problem, and indeed can provide useful insights into variability and opportunities for change. For example, if you are interested in sustainable transport, you might find that almost all actors own fossil fuelled cars, but some will use more fuel than others. Car owners will all have driving skills, but some may be more efficient drivers than others. All might drive their cars regularly, but some may drive more frequently than others. From a high-level perspective, they all share a similar mobility culture—one that is dependent on cars and fossil fuels—but if you drilled down you could identify variations in that culture. Where you choose to place your inclusion–exclusion delineation around this group, and whether you choose to segment it into sub-groups, will depend on the purpose of your research.

Ultimately, it doesn't pay to agonise too much about exactly where to draw a line around the actors and cultural ensembles to study. What we're interested in as researchers is finding patterns that reflect general similarities in cultural features which relate to sustainability outcomes. It is about sense-making through identifying fuzzy patterns of similarity and difference, which is more fruitful than assuming that everyone is identical.

Determining Cultural Vectors

If you are interested in how culture is transmitted, learned and adopted, you may also wish to examine the role of cultural vectors (Fig. 8.3). As discussed in Chapters 4 and 6, vectors include such things as sensory encounters, forms of communication, bodily learning and semantic knowledge that are absorbed from sources such as social interactions, media, bodily experiences and formal learning. Through cultural vectors, people come to adopt similar activities to others, and/or acquire or make similar material items, and/or develop similar norms, aspirations, understandings and other motivators. Vectors are how people learn culture, how it is socially reinforced, and how new cultural concepts are passed from actor to actor.

Cultural vectors will not necessarily be important for all research, but they can help reveal processes of cultural continuity and cultural change. In New Zealand research, for example, we asked householders about alterations they had recently made to improve the heating in their homes, and found that family and friends were by far the biggest influence on their decision. Hearing others' stories of change and experiencing the warmth of others' homes were far more influential than information campaigns, online information or advisory services.

Determining the Agency Boundary

Culture is most often used to describe shared characteristics across a population, but the cultures framework asks researchers to identify a subset of cultural features: those that are both particular to their chosen actors and that could potentially be changed by those actors. This demarcation is indicated in the cultures framework by the agency boundary, shown as a dashed circle around the core elements of culture (Fig. 8.4).

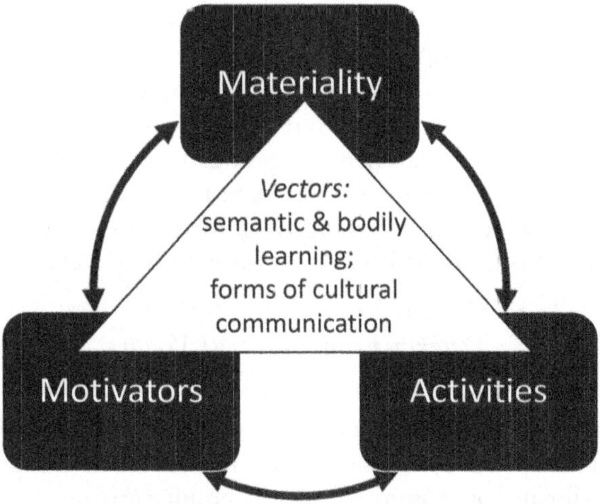

Fig. 8.3 Cultural vectors—the means by which culture is learned and shared

The boundary reflects the capacity of the actor to make choices regarding their cultural ensemble. It distinguishes between elements of culture that are particular to and/or controlled by the actor group under study and those that are particular to and/or controlled by others. The actors' capacity may be constrained by many things, such as their financial circumstances, their age or gender, their education or their familiarity with bureaucratic systems.

In Chapters 4–6, I describe several examples of how this agency distinction is made and used in research. One example is of people on low incomes living within rental housing; the outcomes of interest are energy consumption and wellbeing. Here, the house and chattels owned by the landlord are not considered part of the tenants' energy culture because tenants have no control over them. The tenants' cultural ensemble comprises the dynamic package of motivators, activities and materiality that they enact within those constraints. As well as being shaped by the landlord, their energy culture will be shaped by additional influences beyond the agency barrier, such as the cost of power, government policies and other external influences. Another example is personal mobility, where household actors' cultural ensembles are strongly shaped and constrained by matters beyond their control, such as urban form,

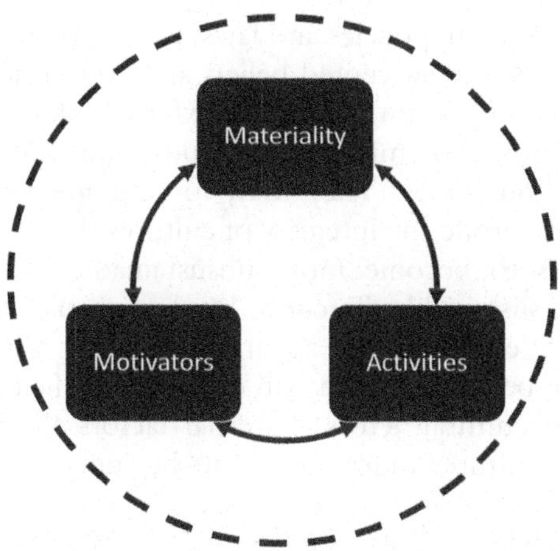

Fig. 8.4 Depicting the agency boundary

the availability of public transport, the safety of walking and cycling and government policies.

The agency boundary in the cultures framework thus invites researchers to differentiate between cultural features that are specific to the group they are studying (within the dashed circle), and other influences that shape (but are rarely influenced by) that culture. It also invites consideration of the factors that are limiting their agency, which may become highly relevant in studies where actors are unable to become more sustainable because of agency constraints. This invites the researcher to consider the implications of differentials in power, the relative responsibility for sustainability outcomes between actors within the agency boundary and those outside it, and their relative ability to act to alter these outcomes. Cultural features beyond actors' agency belong in 'external influences' which may include more powerful cultures.

Determining the External Influences

External influences are exogenous factors that significantly shape the culture that is under study, and are conceptually located outside the agency boundary (Fig. 8.5). As discussed in prior chapters, external influences come in many forms, including qualities of the environment and infrastructure, purposeful policies and laws, pricing regimes, availability of technologies, and broadly accepted beliefs and conventions. For research purposes, it will be important to identify external influences that in some way affect the cultural ensembles under study, and thus ultimately affect the sustainability outcomes. They may, for example, reinforce existing cultural ensembles, erode the integrity of cultures that are already sustainable, force actors to become more unsustainable or support cultural change in a more sustainable direction. Depending on the research focus, some external influences may be apparent from the outset of the study, while others may be obscure and will need to be elicited through deep engagement with cultural actors. External factors that clearly have no influence on the culture under study can be ignored for the purposes of cultural analysis.

External influences can also be interpreted as broader cultures that influence the culture you are focusing on. For example, the mobility cultures of citizens are strongly shaped by cultural features of the municipality. A council that decides to invest its transport funding primarily in new motorways is not doing so arbitrarily. The decision will have emerged from a well-established system of beliefs, understandings, aspirations and organisational practices within the council—in other words, their culture. In researching external influences, it may therefore be important to look

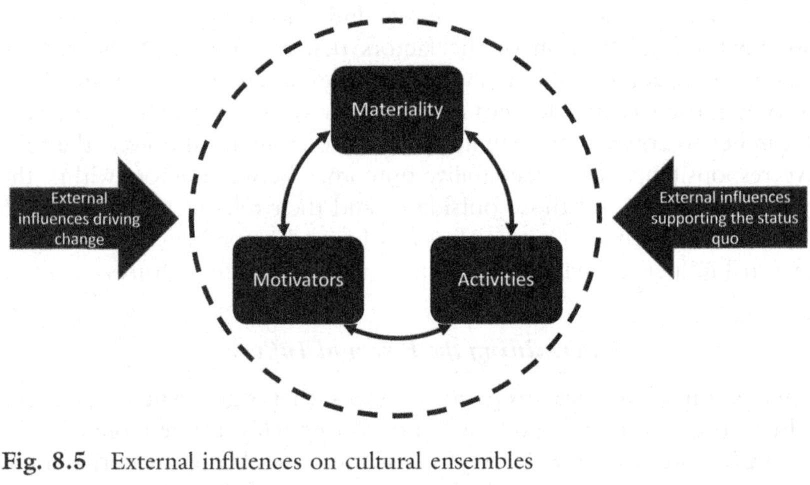

Fig. 8.5 External influences on cultural ensembles

beyond their presenting qualities to understand the cultures within which they are embedded and replicated.

Another external influence to consider is your own impact on culture as a result of the research process. By asking questions of research participants, you are likely to be raising their own awareness of aspects of their culture that they may not have considered previously. In your interactions, even if it is not intended, you may make them more aware of the sustainability outcomes of their cultural ensembles, and/or aware of disjunctions between, say, their beliefs and practices, or between their aspirations and material possessions. Your interactions may also cause them to develop new understandings about the sustainability issue of interest, or open their eyes to external influences that they had not previously been aware were shaping their culture. The research process is never neutral, so be aware of how your work may influence your participants' cultures, ensure that your work is carried out ethically and does no harm, and possibly build an evaluation of your impact as a researcher into your research.

Investigating Cultural Stability

Some cultures change very little over time, or at least in the features that give rise to sustainability outcomes. Some enduring cultural ensembles may be positive examples that research can learn much from, such as communities and organisations that have consciously set out to become more sustainable and have maintained that over time. Other cultural ensembles are deeply problematic from a sustainability perspective, and yet continue to endure. We can see this with highly consumptive lifestyles amongst many in the Western world, with beliefs in the value of consumption for its own sake, aspirations for material items goaded by the media (and today, by social influencers), practices of shopping valued in their own right and made more unsustainable through the proliferation of short-lived products, and wellbeing equated with more (or more wealth-signifying) possessions. The cultures framework offers a structure for exploring how and why many unsustainable cultural ensembles change little over time. It would be helpful to review the examples of research into cultural stability discussed in Chapter 5.

Research into cultural stability might start by examining actors' cultural ensembles, exploring relevant motivators, material assemblages and activities, and the extent to which these are aligned and mutually supportive. It may investigate cultural vectors in order to understand how cultural attributes are learned, reinforced and conveyed to new members. It would usefully identify what external influences (including structures, institutions and broader-scale cultures) are supporting the culture in its current form and enabling its continuance. It may be useful to look back in time and identify what has shaped the culture you see today. Unpacking these dynamic interactions can help explain how and why this culture is resistant to change.

Understanding these dynamics is particularly important if your research is seeking to understand why change interventions have failed to achieve their targets. Exploring culture as a dynamic system can reveal why change in one external influence or in a single cultural feature (e.g. new knowledge or a new technology) may make little or no difference to sustainability outcomes; for example because its impact is moderated by other cultural dynamics that tend to stabilise the cultural ensemble. Even if some aspects of culture (e.g. values, aspirations) are aligned with sustainable outcomes, other aspects (e.g. routines, agency limitations, external influences) may prevent or limit overall change. It is critical to gain a better understanding of the dynamics of cultural stability if we are to achieve widespread sustainability transitions.

Investigating Cultural Change

The cultures framework can also underpin research on how cultures change. From a sustainability perspective, your investigation might be into positive change, such as new cultural features being adopted with cascading impacts on the entire cultural ensemble. Of particular interest here might be how positive change processes are initiated, and the consequential effects on culture. An example that I described in Chapter 6 was how the replacement of kerosene lamps with solar lights in Vanuatu had a domino effect on many other aspects of culture including everyday practices, gender roles, beliefs, and aspirations for other solar and digital technologies. Your study could equally focus on negative cultural change, seeking to understand how sustainability-oriented cultural characteristics have been lost. For example, in Chapter 6 I described a Māori community where degradation of the inshore fisheries meant that community members could no longer gather traditional foods. The inability to undertake practices resulted in a loss of knowledge and skills that had previously sustained the fishery.

With the cultures framework as a structuring device, a researcher can explore what external influences might be tending to encourage change, as well as what changes are already occurring within that culture and whether these are leading to shifts in other cultural features. In the previous section, I discussed how cultural ensembles could become resistant to change due to strong alignments between motivators, activities and materiality. In contrast, systems where there are misalignments (e.g. aspirations are different to practices; material items don't fit with beliefs) there is a greater potential for instability, innovation and change. Researchers interested in the potential for change might wish to examine the degree to which the relevant cultural elements are aligned.

Studying the processes of cultural change is critically important to sustainability transformations at all scales. The cultures framework can help to systematise analysis of where cultural change starts, whether it leads to consequential change to the cultural ensemble, the sustainability consequences of this change, and whether incipient changes are prevented by other factors. Cultural change is unlikely to occur all at once—it may involve incremental adjustments to the ensemble over time (e.g. a normative shift may precede a behavioural shift, or a new technology may precipitate new practices). Change also will not be uniform across a culture group, so the analysis may need to include identification of which actors have first made these changes, and through what vectors this has become more widely adopted. More research is needed to better understand the uneven, incremental processes of cultural change as well as the circumstances in which rapid transformation can occur.

People within a culture group rarely get to alter the more powerful external influences shaping their culture. But sometimes it happens, and this is possibly the most powerful driver of transformational change, as discussed towards the end of Chapter 6. This is where cultural changes spread widely across less powerful actors, and membership of the new culture group grows to the extent that it starts to have influence beyond the agency barrier, reshaping the motivators, activities and/or materiality of more powerful actors. If researchers are interested in the potential for radical sustainability transformations, they should focus on the potential for outward as well as inwards flows through the agency barrier. The urgency of the sustainability crisis means that we need to know as much as possible about how to achieve rapid and widespread transformations of dominant unsustainable cultures.

Having an Impact

By now, your research will have produced an understanding of the various elements of the cultures framework and how they interact dynamically, and any external influences that are tending to prevent or enable change. You will understand the limits of actors' agency, sub-cultures may have been identified and you may have also discovered cultural influences at other scales. You will have gathered evidence as to whether the cultural ensemble has positive or negative outcomes for the relevant measures of sustainability. You will know whether the culture is in the process of change or is resilient to change, and why this may be the case. If the ensemble has poor sustainability outcomes, your findings should indicate whether it has some latent potential for more sustainable change and possible ways in which change could be initiated or supported to achieve more sustainable consequences.

As a researcher, you might want to apply your findings further to actively help in the sustainability transition. Does it show a culture that already has great sustainability outcomes? If so, what can we learn from this and how can this success be supported and amplified? Does the research show a culture that is stuck in unsustainable patterns and unable to change? If so, where are the opportunities to support change? Does it show a culture that is gradually becoming more sustainable but has a way to go? If so, how can that journey be supported? Does is show a culture where attempts have been made towards greater sustainability but those attempts have been unsuccessful? If so, how can your findings help show why this might be the case?

As with many of the research projects using the cultures framework, you could develop recommendations for policy or practical actions. You might build on your work and develop a programme of action research that enables your insights to be trialled. You might assist an organisation or group of actors to interrogate their own culture and begin a process of change. You could collaborate with an already sustainable community over how to challenge the forces that are depleting it, or how to use cultural vectors to extend its reach. There are endless possibilities for making cultural research into a force for positive change.

Research Methodologies

Research with the cultures framework can be undertaken using a broad sweep of qualitative and quantitative methodologies. In this section, I outline the range of research methods that have been successfully used to date that I am aware of, and what functions these methods have played. This section is heavily referenced so that readers can go to the original papers for more detail on the specific methods of data elicitation and analysis.

Most studies to date using the cultures framework have used qualitative methods to examine cultural ensembles, either on their own or in combination with quantitative methods. Solely qualitative research often involves interviews followed by analysis to draw out evidence illustrating cultural elements (e.g. Bach et al., 2020; Lazowski et al., 2018; McKague et al., 2016; Scott & Lawson, 2018; Tesfamichael et al., 2020; Walton et al., 2014). Some projects have used a combination of qualitative methods such as workshops or focus groups together with interviews (e.g. Ambrosio-Albalá et al., 2019; Godbolt, 2015; Krietemeyer et al., 2021). A study of cultural change over an extensive period of time incorporated reviews of archaeological and historical evidence together with present-day interviews (Stovall, 2021). Researchers often apply thematic analyses to their qualitative material, but other analytical methods can be used. For example, a study on energy cultures of poverty analysed the interview texts using a computational social science methodology (Debnath et al., 2021).

Other researchers have used quantitative methods to characterise cultural ensembles. The elements of the framework have underpinned the design of surveys to elicit data from a larger population than is possible with face-to-face qualitative methods (e.g. Lawson & Williams, 2012) and as the basis of an 'energy culture' survey for businesses to determine the maturity of their energy efficiency efforts (Oksman et al., 2021). As well as using data produced from surveys specifically designed for this purpose, the cultures framework has been used retrospectively to underpin analysis of existing data sets to identify clusters of similar cultural characteristics aligned with different sustainability outcomes (e.g. Bardazzi & Pazienza, 2017, 2018, 2020; Walton et al., 2020). It has also been used as an integrating framework across multiple quantitative data sets (e.g. Manouseli et al., 2018). There will always be variations in the cultural ensembles of any group of actors, and sometimes it will be useful to explore this variability. Larger quantitative data sets have been used as a basis for segmenting populations into statistically distinctive groups using cluster analysis (e.g. Barton et al., 2013; Lawson & Williams, 2012).

In studies using mixed qualitative and quantitative methods, the cultures framework has underpinned both design and/or analysis, and has been used to facilitate the integration of findings (e.g. Bell et al., 2014; Muza & Thomas, 2022; Scott et al., 2016). Some studies have gathered both qualitative and quantitative data relating to cultural elements during face-to-face interviews and integrate these in the analysis (e.g. Khan et al., 2021). Research on indicators of national energy cultures used a combination of policy analysis and quantitative analysis of comparative data sets (Stephenson et al., 2021).

To explore external influences (including multi-level cultures), studies often ask interviewees within the culture group about their perceptions of what shapes their decision-making or constrains their ability to make more sustainable choices (e.g. Ambrosio-Albalá et al., 2019; Debnath et al., 2021; McKague et al., 2016). Some also seek the views of experts or key informants in particular fields (e.g. Stephenson et al., 2015) or review the impact of laws and policies (e.g. Barton et al., 2013). Some projects have also interviewed actors who represent aspects of external influences (e.g. Jürisoo et al., 2019; Nicholas, 2021).

Many different research approaches can be used to identify causal relationships between cultural ensembles and sustainability outcomes. Some studies have done this quantitatively, such as identifying relationships between householder age cohorts and energy consumption (Bardazzi & Pazienza, 2017) and between timber drying cultures and energy use (Bell et al., 2014). One study used regression analysis to relate householders' cultural features to their interest in being involved in a local energy management scheme (Krietemeyer et al., 2021). However, in most studies to date using the cultures framework causal relationships are not quantified but are assumed based on well-established understandings of sustainable practices or the impacts of different technologies (e.g. Dew et al., 2017; Hopkins, 2017). Often the focus of research has been on whether the cultural ensemble has features that are known to align with more sustainable outcomes (e.g. types of technology and practices that represent business energy efficiency [Oksman et al., 2021; Walton et al., 2020]) rather than setting out to prove the well-understood relationship between these and measurable outcomes.

The framework has also assisted with modelling. A design for agent-based modelling for smart grid development drew from the cultures framework to incorporate energy use behaviours into the models (Snape et al., 2011). A project using system dynamics modelling of the uptake of electric vehicles also used the elements of the cultures framework as foundational data for the model (Rees, 2015). Methods such as these align well with the original conceptual framing of culture as a system, and offer a dynamic structure to explore system-type interactions between components of the framework.

The framework lends itself to multi-scalar analysis, as with research on PV uptake in Switzerland, where the work described generalised cultural ensembles of adopters and non-adopters and also drew insights on cultural processes from individuals (Bach et al., 2020). By focusing on collectives, researchers can observe patterns of similar cultural features across a population and identify broadly similar influences on and outcomes of that culture. By focusing on individual actors, they can also explore in detail the dynamics within cultural ensembles.

The framework has also been used to structure reviews of literature. Examples include reviews of the adoption of energy-efficient technology innovations in buildings (Soorige et al., 2022), academic air travel cultures (Tseng et al., 2022) and adoption of natural gas (Binney & Grigg, 2020). In a study on barriers and drivers for industrial energy efficiency, the framework was refined to fit an industrial context and used as an organising framework for metadata from a literature review on the barriers and drivers of energy behaviour in firms. This approach enabled the researchers to consider many interdependent components of efficiency decision-making by industry, including attitudinal factors, behaviours and technologies (Rotzek et al., 2018).

Some studies have investigated the effectiveness of interventions intended to improve sustainability outcomes. In these instances, they have used the framework to guide collection of pre-intervention and post-intervention data on cultural ensembles and/or outcomes (e.g. Rau et al., 2020; Scott et al., 2016). Other work has used the framework to design evaluation tools (e.g. Ford et al., 2016; Karlin et al., 2015).

Within larger research programmes, the framework can be used as a structuring device for allocating research roles and methods across an interdisciplinary team. In the Energy Cultures 1 and 2 research programmes, for example, we identified the core elements of culture in relation to energy efficiency (material aspects, practices, norms, beliefs, etc.) through householder questionnaires, focus groups and interviews. To relate cultural ensembles to energy outcomes, we included questions about energy consumption in the surveys and in later work we used data from smart electricity meters. Interrelationships between elements of the framework were explored in various ways. We used a values 'laddering' approach (from consumer psychology) to look at the relationships between values and household energy efficiency actions (Mirosa et al., 2013) and used choice modelling (an economics tool) to examine the interactivity between people's motivators and preferences for adoption of efficient technologies (Thorsnes et al., 2017). These were staged so that the values work helped in the design of the choice modelling. We explored interactions between norms, material culture and practices with community focus groups, and these groups also assisted in identifying external influence that were barriers to changing behaviour. Desktop studies were used to examine regulatory, market and policy influences on energy culture. We used social network analysis to identify the most common sources of external influence on householder choices to adopt more efficient technologies. All of these different sources of insight were linked though the framework, which supported an integrative approach across the team, learning from each other's findings and contributing to a holistic understanding of household energy cultures in the New Zealand context (Barton et al., 2013; Stephenson et al., 2010).

The framework is thus helpful in underpinning the design of research as a single- or multi-method project by an individual researcher, or a multi-method multidisciplinary research programme by a team. It can be used proactively to design research and analyse the findings, or used retrospectively to help analyse existing data from single or multiple sources. It is fruitful when used as a theory in its own right, and also when used in combination with other theories.

Using the Cultures Framework as a Meta-Theoretical Framing

As these examples have shown, research using the cultures framework is not confined to particular methods, and neither is it confined to any particular theoretical or disciplinary perspective. In this sense, the framework offers a meta-theoretical set of universal elements, and leaves it to the researcher to determine which theories and methodologies are best used to examine them. Rau et al. (2020) describe the advantages of the framework thus:

> The benefits of using the [cultures framework] to organise the empirical material and findings of this interdisciplinary energy research were considerable, especially given its focus on the multi-method investigation of a small number of households. Its relative simplicity, easy-to-understand terminology and focus on both social and material aspects of energy use made it an ideal tool for fusing insights from the social sciences, engineering and architecture. At the same time, [the framework] was capable of connecting a higher-order theoretical approach (energy cultures) to concrete empirical energy-related outcomes. (p. 10)

While many studies use the cultures framework as a framing theory in its own right, it is at the same time an organising framework that enables multiple methods, theories and disciplines to contribute to an understanding of culture in relation to sustainability. Studies using the cultures framework to date have drawn from complementary explanatory theories as diverse as sociological theories of agency, structure, institutions and practice, theories of power and gender, behavioural theories, socio-technical systems theories, consumer psychology, economic theories, the multi-level perspective and, of course, theories of culture. Generally, these are used to inform analysis of an aspect of the cultures framework.

This flexibility is well explained by Ambrosio-Albalá et al. (2019) in their conclusion to a paper on public perceptions on distributed energy storage in the United Kingdom:

> … we find that the framework functions as a useful heuristic, allowing us to organise and reflect on a wide range of factors in a way that is more inclusive than a psychology-only perspective. The idea of there being multiple possible cultures in relation to energy use – and the observation of these at different scales – also helps to stimulate thinking on further research directions in terms of how different households, demographic segments, nationalities and entities may differ in terms of the nexus of norms, attitudes, behaviours or practices and material experiences. These cultures will likely need different types of communication, informational, institutional and contractual offers, given likely differing responses. A further value of the ECF [energy cultures framework] – regarding which we would concur with its originators – lies in its comprehensibility for non-social scientists. For more specialised and narrowly specified forms of analysis, we would

> defer to the psychological and sociological perspectives that the ECF draws upon. (p. 149)

There are many under-explored possibilities for the use of other theories and bodies of knowledge to help explore aspects of culture in relation to sustainability. For example, in relation to the theory of structuration, culture works both to replicate social life and as a creative force for change. The framework positions culture as constrained and shaped by structure, while simultaneously situating more powerful cultures as part of structure. Despite these constraints on their agency, cultural actors can and do make independent choices and can collectively reshape more powerful structures and cultures. This interplay (and the conceptual overlap of culture and structure) invites further exploration in both a theoretical and applied sense, particularly in the context of the implications for sustainability transitions.

Conceptual fields that underpinned the development of the cultures framework could be drawn from more extensively, including lifestyles literatures, socio-technical studies, actor network theory, systems approaches, and sociological and anthropological theories of culture. For example, social practice theory can help illuminate aspects of the inner elements of the framework, with a focus on habitual actions. Theories of power and justice can help elaborate on the reasons for limitations in agency and choice that are imposed by those outside the agency boundary. Socio-technical systems theories can assist in exploring the relationships between actors' material items and their activities. Theories of gender can help explores difference in cultural meaning, gender equity and gender leadership in sustainability outcomes. In all of these ways and more, the framework can offer a meta-theoretical structure for deeper analysis depending on the inclinations and interests of the researcher.

Further Contributions From Cultural Theory

A further untapped potential lies in the application of cultural theories more generally to questions of sustainability. Cultural theory is a vast field that I could only sketch out lightly in Chapters 2 and 3. There I discussed divergences and similarities across cultural theories and identified nine main clusters of perspectives on culture. I believe that each of these perspectives on culture can make an important contribution to research for a more sustainable future.

Culture-as-nature is the oldest of the nine perspectives. It is mostly overlooked by dominant ideologies, and yet its endurance offers the most hope. Culture-as-nature reflects many Indigenous perspectives that defy the intellectual separation of human society and natural systems. Culture-as-nature recognises our utter dependence on the natural world. The most powerful expressions of culture-as-nature continue to come from Indigenous peoples, although recent years have seen an increasingly strong voice from Western scholars (e.g. Haraway, 2016; Plumwood, 2005; Tsing et al., 2017). Culture-as-nature reinforces the indivisibility of human existence from nature and the responsibilities of human societies to maintain the integrity of natural systems. It also breaks down the barriers of cultural membership. Natural features are actors in culture: mountains and creatures are family members, trees communicate, rivers are people; they are all cultural members with agency. Many of the Indigenous societies of the world offer principles, values, practices, knowledges and worldviews that are crucial for a sustainable future (Artelle et al., 2018; Mazzocchi, 2020; Watene & Yap, 2015; Waldmüller et al., 2022; Yunkaporta, 2020).

Culture-as-nurture reflects the original meaning of culture in old English, referring to processes of husbandry—the careful tending of crops and animals. For the sustainability crisis, we are relearning the urgency of nurturing all life forms and regenerating natural systems. As well as reinforcing the importance of healthy natural systems and food production, culture-as-nurture can be further interpreted as the re-grounding of communities in caring for place, and reviving the spiritual roots of agriculture (Bisht & Rana, 2020; van den Berg et al., 2018). Urban agriculture or community gardens similarly reconnect people to the practices and rhythms of caring for nature, along with the sharing of food and strengthening a sense of community (Sumner et al., 2011). Culture-as-nurture reflects the ways in which we must re-learn practices of caring for nature, and how caring for nature aligns with caring for each other.

The original sense of culture-as-progress was the process of human development towards a so-called civilised state that reflected certain Western ideals of art and behaviours. Although it is now repellent and largely obsolete in this original sense, the idea of 'progress' can be reconfigured to refer to cultural journeys towards sustainability. Culture-as-progress in this sense can recognise the many cultural configurations that already have sustainable outcomes. It invites investigations of factors that underpin the relative sustainability of one culture compared to another (Buenstorf & Cordes, 2008; Minton et al., 2018) and the application of cultural evolution concepts to sustainability challenges (Brooks et al., 2018). A practical application of this in the world of business is the concept of energy culture 'maturity' and evaluation methods to assess such progress (Soorige et al., 2022). If the concept of progress is applied to outcomes rather than to cultural characteristics in themselves, this removes the suggestion that certain forms of cultural ensemble are better than others. Instead, a sustainable future requires a multitude of sustainable cultural ensembles specific to people and place, at a multitude of scales.

Culture is still commonly used to refer to works and practices of artistic and intellectual activity. In this sense, culture-as-product plays an important role as a cultural vector in transmitting ideas, values and possibilities. For the sustainability transition, creative works will play a critical role in challenging systems, institutions and practices that are destroying natural systems and demeaning humanity, as well as offering inspiration for alternate futures. Cultural products have the potential to convey different understandings, such as about the world's ecological limits, actions for sustainability and new perspectives of the future (Curtis et al., 2014). This is already a strong theme in art and performance (Galafassi et al., 2018; Kagan, 2019) but could play an even stronger role in helping shape awareness and collective visioning for a sustainable future.

Culture-as-lifeways draws originally from anthropological studies of the distinctive way of life of a group of people. From a sustainability perspective, this concept can be redirected from studying the ways of life as a focus in their own right, to looking at the relationship between ways of life and the sustainability outcomes. From a research perspective, it encourages work that explores the variety of ways that people already live sustainably—for example, differences between ways of life in the global north and global south (Hayward & Roy, 2019), as well as how group or community ways of life can be re-oriented towards more sustainable consequences (Brightman & Lewis, 2017).

Culture-as-meaning focuses on the shared meanings and symbolisms of cultural objects such as text, discourse and possessions. For the sustainability challenge, culture-as-meaning can help reveal the ways in which symbolism can work for or against sustainable outcomes. Theories of cultural meaning could be applied to the analysis of how unsustainability is inherent in dominant rhetoric, text and discourse (e.g. Sturgeon, 2009), and the ways in which new meanings are being forged as part of cultural transformations (e.g. Hammond, 2019). Other examples of work using culture-as-meaning include a study of how the term 'sustainable development' is constructed in the disclosures of Finnish-listed companies (Laine, 2005), how the media interprets sustainability (Fischer et al., 2017) and the importance of symbolism in marketing for the sustainability transition (Kumar et al., 2012; Sheth & Parvatiyar, 2021).

Culture-as-structure is interested in the underlying rules by which social systems are reproduced—the cultural codes of social life. Culture-as-structure, as embedded in institutions and discourses, is intimately tied with questions of power and influence regardless of intent (Blythe et al., 2018). Drawing on this literature can help identify and challenge ideologies, assumptions and rules that replicate unsustainability. Relevant studies using culture-as-structure include how neoliberal ideology operates through sustainability discourses (Jacobsson, 2019) and the mental structures in which finance actors are embedded (Lagoarde-Segot & Paranque, 2018).

Culture-as-practice studies the bodily practices that produce and replicate social life, and the intimate linkages between routines, objects, meanings and competencies. This field of work can contribute to questions of how to alter practices, or develop new practices that support sustainability. Practice theory has already been widely applied to how to achieve less resource-intensive habits and routines, including how the reproduction of social practice can sustain inequality and injustice (Shove & Spurling, 2013) and to provide insights on collective action for social change (Welch & Yates, 2018).

Culture-as-purpose reflects bodies of work that focus on how to change the culture of organisations or groups of actors intentionally. Work in the field of organisational culture includes how to deliberately create more sustainable organisations (Galpin et al., 2015; Obal et al., 2020). Education for sustainability is another major field working on purposeful culture change, building on and extending educational theories (Huckle & Sterling, 1996) and education's transformative potential (Filho et al., 2018). This includes using practices of dance and music to develop pro-social behaviours that align with sustainability goals (Bojner et al., 2022).

All nine conceptualisations of culture thus make important contributions to understanding the role of culture in sustainability. It is evident that at least some academics in each of these fields are applying relevant theories and methodologies to sustainability questions, but it appears to be occurring in a fragmented way with different bodies of knowledge scarcely acknowledging each other. Even if there is a ramping up of scholarly contributions on culture and sustainability, there is the risk that the slipperiness of culture as a concept will continue to handicap the use of research findings by practitioners and policymakers. If culture continues to be presented as if each part of the elephant is the full elephant, and the only true elephant, its ongoing indeterminacy will continue to confuse potential research users and dilute the effectiveness of scholarly contributions.

The cultures framework could help here by 'locating' these different approaches to culture and their contribution to sustainability challenges. Each of the nine clusters of meaning can offer insights for certain features or qualities of the framework. Using its meta framing, culture-as-purpose focuses on how to purposefully initiate cultural change and achieve better sustainability outcomes. Culture-as-practice focuses on routines as a subset of activities, emphasising how they cannot be understood in isolation from the objects, competencies and meanings associated with them. Culture-as-structure helps explore entrenched external influences or higher-order cultures that use their power to shape the cultural ensembles of less powerful actors. Culture-as-meaning focuses on the meanings and symbolism of activities and objects and can help illuminate the mechanisms of cultural vectors.

Culture-as-lifeways scholarship can help in studying how culture is learned, the dynamics of the core elements of culture, processes of replication or change, and the heterogeneity of cultural ensembles. In the arts, culture-as-product scholarship can help enhance the role of creative activities and products in building a more sustainable future. Academic fields that align with culture-as-progress focus more on the two-ended arrow that links cultural ensembles and outcomes and can help with studies on the many journeys involved in achieving cultures that touch lightly on the earth.

Culture-as-nurture scholarship contributes to the adoption of more nurturing food-production activities that also enhance social and environmental outcomes. Culture-as-nature opens the door to entirely different worldviews and knowledge systems regarding humans' relationships with the natural world. Importantly, it extends cultural actors to include non-human life forms, spiritual beings and landscape features. In this way, it offers ways of understanding the sustainability transition as a process of restoring health and vitality to all living things and the natural systems that support them.

The cultures framework can thus work as an integrative heuristic, indicating how different interpretations of culture all contribute in important ways to a fuller understanding of the role of culture in sustainability. Used in this way, the framework can help reveal the complementary roles of these diverse approaches to practitioners and non-cultural academics. It can also indicate to researchers where it might be useful to reach back to the original bodies of cultural theory and bodies of knowledge to help illuminate particular aspects of the overall 'elephant of culture'. In this way, cultural scholarship can be used more comprehensively and systematically to support sustainability transitions, and the slipperiness of culture is somewhat reduced.

Conclusion

As this chapter has demonstrated, there is no 'right way' to do research with the cultures framework. It is a set of highly generalised variables and their relationships, which can be explored using a wide range of research methods and theories. Researchers can choose which features within the general variables to focus on and can apply the framework at any scale and to any set of actors. The framework can be used as a framing theory in its own right, or alternatively (or simultaneously) can operate as an integrating frame for interdisciplinary, multi-method research, or as a meta-theoretical framing of the complex field of culture and sustainability. Either on its own or in combination with other frameworks and theories, the cultures framework thus offers ontological and epistemological inclusiveness for transdisciplinary research agendas.

There is an urgent need for a better understanding of the role of culture in the sustainability crisis and how to transform the unsustainable cultures that are inherent in most systems of production and consumption. Technologies alone will not achieve a net-zero world by 2050, or turn around our devastating losses of biodiversity, or enable equitable access to energy for families in developing countries. It will take more than simply changes in behaviour. It will require fundamental changes in the motivators, activities and materialities of people and organisations at every scale. We can learn much from cultures that are already sustainable or are on journeys of transition, but the biggest challenge is how to achieve transformational cultural change, at scale, and with unprecedented speed. To that end, research is desperately needed to improve our understanding of processes of cultural change, and particularly to understand how powerful meta-cultures can be destabilised and their unsustainable ideologies and institutions transformed.

Although the framework has already been used in a wide variety of fields, it has the potential to do much more to assist with journeys of transition. We need to know more about how sustainable cultures develop and endure, a better understanding of the dynamics of cultural change and the role of vectors in cultural learning, as well as processes of cultural expansion and collectivisation. More research is needed on the implications of actors having multiple cultural ensembles in different aspects of their life (e.g. the interplay between their food culture, mobility culture and household energy culture). Researchers could fruitfully explore how these overlapping cultures influence and shape each other, or divergences between cultural settings at home and at work. We need to better understand how cultural transformations can be enabled, and the roles that culture will need to play to achieve sustainability transitions. And while some research using the framework has explored issues of power and justice, there is much more to be done.

Cultural research can work at two levels in this quest for transformation. On the one hand, cultures are unique as to membership and cultural features. Researchers can draw conclusions regarding specific cultural ensembles and their sustainability outcomes, and can make recommendations for change initiatives. However, these will usually only be relevant to the case in question. On the other hand, as more studies are undertaken, we can start to build up generalisable understandings of cultural dynamics as they relate to sustainability. Across multiple studies, the research community can develop a better picture of universal cultural processes and effective change interventions.

Finally, although this chapter is about culture and sustainability, the research approach I describe in this chapter could be used in other fields of inquiry. It was developed for sustainability-related research and has mainly been used for that purpose, but it could equally be applied to investigations of culture for any other reason, and in relation to any other outcome. But my hope is that it will continue to be primarily used for research that helps achieve a just transition to a sustainable future.

References

Ambrosio-Albalá, P., Upham, P., & Bale, C. S. E. (2019). Purely ornamental? Public perceptions of distributed energy storage in the United Kingdom. *Energy Research & Social Science, 48*, 139–150.

Artelle, K. A., Stephenson, J., Bragg, C., Housty, J. A., Housty, W. G., Kawharu, M., & Turner, N. J. (2018). Values-led management: The guidance of place-based values in environmental relationships of the past, present, and future. *Ecology and Society*, *23*(3).

Bach, L., Hopkins, D., & Stephenson, J. (2020). Solar electricity cultures: Household adoption dynamics and energy policy in Switzerland. *Energy Research and Social Science*, *63*, 101395.

Bardazzi, R., & Pazienza, M. G. (2017). Switch off the light, please! Energy use, aging population and consumption habits. *Energy Economics, 65*, 161–171.

Bardazzi, R., & Pazienza, M. G. (2018). Ageing and private transport fuel expenditure: Do generations matter? *Energy Policy, 117*, 396–405.

Bardazzi, R., & Pazienza, M. G. (2020). When I was your age: Generational effects on long-run residential energy consumption in Italy. *Energy Research & Social Science, 70*, 101611.

Barton, B., Blackwell, S., Carrington, G., Ford, R., Lawson, R., Stephenson, J., Thorsnes, P., & Williams, J. (2013). *Energy cultures: Implication for policymakers*. Centre for Sustainability, University of Otago.

Bell, M., Carrington, G., Lawson, R., & Stephenson, J. (2014). Socio-technical barriers to the use of energy-efficient timber drying technology in New Zealand. *Energy Policy, 67*, 747–755.

Binney, M., & Grigg, N. (2020). The energy cultures framework: Understanding material culture barriers to the uptake of natural gas. *Journal of Asia Entrepreneurship and Sustainability, 16*(5), 105–139.

Bisht, I. S., & Rana, J. C. (2020). Reviving the spiritual roots of agriculture for sustainability in farming and food systems: Lessons learned from peasant farming of Uttarakhand hills in North-western India. *American Journal of Food and Nutrition, 8*(1), 12–15.

Blythe, J., Silver, J., Evans, L., Armitage, D., Bennett, N. J., Moore, M., Morrison, T. H., & Brown, K. (2018). The dark side of transformation: Latent risks in contemporary sustainability discourse. *Antipode*, *50*(5), 1206–1223.

Bojner Horwitz, E., Korošec, K., & Theorell, T. (2022). Can dance and music make the transition to a sustainable society more feasible? *Behavioral Sciences, 12*(11), 1–15.

Brightman, M., & Lewis, J. (Eds.). (2017). *The anthropology of sustainability: Beyond development and progress*. Palgrave Macmillan.

Brooks, J. S., Waring, T. M., Borgerhoff Mulder, M., & Richerson, P. J. (2018). Applying cultural evolution to sustainability challenges: An introduction to the special issue. *Sustainability Science, 13*(1), 1–8.

Buenstorf, G., & Cordes, C. (2008). Can sustainable consumption be learned? A model of cultural evolution. *Ecological Economics, 67*(4), 646–657.

Curtis, D. J., Reid, N., & Reeve, I. (2014). Towards ecological sustainability: Observations on the role of the arts. *SAPIENS: Surveys and Perspectives Integrating Environment and Society, 7*(1), [online].

Debnath, R., Bardhan, R., Darby, S., Mohaddes, K., Sunikka-Blank, M., Coelho, A. C. V., & Isa, A. (2021). Words against injustices: A deep narrative analysis of energy cultures in poverty of Abuja, Mumbai and Rio de Janeiro. *Energy Research & Social Science, 72*, 101892.

Dew, N., Aten, K., & Ferrer, G. (2017). How many admirals does it take to change a light bulb? Organizational innovation, energy efficiency, and the United States Navy's battle over LED lighting. *Energy Research & Social Science, 27*, 57–67.

Filho, W. L., Raath, S., Lazzarini, B., Vargas, V. R., Souza, L. De, Anholon, R., Quelhas, O. L. G., Haddad, R., Klavins, M., & Orlovic, V. L. (2018). The role of transformation in learning and education for sustainability. *Journal of Cleaner Production, 199*, 286–295.

Fischer, D., Haucke, F., & Sundermann, A. (2017). What does the media mean by 'sustainability' or 'sustainable development'? an empirical analysis of sustainability terminology in german newspapers over two decades. *Sustainable Development, 25*(6), 610–624.

Ford, R., Karlin, B., & Frantz, C. (2016). *Evaluating energy culture: Identifying and validating measures for behaviour-based energy interventions*. Proceedings of the International Energy Policies & Programmes Evaluation Conference, 1–12.

Galafassi, D., Kagan, S., Milkoreit, M., Heras, M., Bilodeau, C., Bourke, S. J., Merrie, A., Guerrero, L., Pétursdóttir, G., & Tàbara, J. D. (2018). 'Raising the temperature': The arts in a warming planet. *Current Opinion in Environmental Sustainability, 31*(2), 71–79.

Galpin, T., Whittington, J. L., & Bell, G. (2015). Is your sustainability strategy sustainable? Creating a culture of sustainability. *Corporate Governance, 15*(1), 1–17.

Godbolt, Å. L. (2015). The ethos of energy efficiency: Framing consumer considerations in Norway. *Energy Research & Social Science, 8*, 24–31.

Hammond, M. (2019). Sustainability as a cultural transformation: The role of deliberative democracy. *Environmental Politics, 29*(1), 173–192.

Haraway, D. (2016). *Staying with the trouble: Making kin in the Chthulucene*. Duke University Press.

Hayward, B., & Roy, J. (2019). Sustainable living: Bridging the north-south divide in lifestyles and consumption debates. *Annual Review of Environment and Resources, 44*, 157–175.

Hopkins, D. (2017). Destabilising automobility? The emergent mobilities of generation Y. *Ambio, 46*(3), 371–383.

Huckle, J., & Sterling, S. R. (Eds.). (1996). *Education for sustainability*. Earthscan.

Jacobsson, D. (2019). In the name of (Un)sustainability: A critical analysis of how neoliberal ideology operates through discourses about sustainable progress and equality. *TripleC, 17*(1), 19–37.

Jürisoo, M., Serenje, N., Mwila, F., Lambe, F., & Osborne, M. (2019). Old habits die hard: Using the energy cultures framework to understand drivers of household-level energy transitions in urban Zambia. *Energy Research and Social Science, 53*, 59–67.

Kagan, S. (2019). Culture and the arts in sustainable development. In T. Meireis & G. Rippl (Eds.), *Cultural Sustainability: Perspectives from the humanities and social sciences* (pp. 127–139). Routledge.

Karlin, B., Ford, R., & McPherson Frantz, C. (2015). Exploring deep savings: A toolkit for assessing behavior-based energy interventions. *International Energy Program Evaluation Conference*, 1–12.

Khan, I., Jack, M. W., & Stephenson, J. (2021). Dominant factors for targeted demand side management: An alternate approach for residential demand profiling in developing countries. *Sustainable Cities and Society, 67*, 102693.

Krietemeyer, B., Dedrick, J., Sabaghian, E., & Rakha, T. (2021). Managing the duck curve: Energy culture and participation in local energy management programs in the United States. *Energy Research & Social Science*, 102055.

Kumar, V., Rahman, Z., Kazmi, A. A., & Goyal, P. (2012). Evolution of sustainability as marketing strategy: Beginning of new era. *Procedia—Social and Behavioral Sciences, 37*, 482–489.

Lagoarde-Segot, T., & Paranque, B. (2018). Finance and sustainability: From ideology to utopia. *International Review of Financial Analysis, 55*, 80–92.

Laine, M. (2005). Meanings of the term "sustainable development" in Finnish corporate disclosures. *Accounting Forum, 29*(4), 395–413.

Lawson, R., & Williams, J. (2012). *Understanding energy cultures*. Annual Conference of the Australia and New Zealand Academy of Marketing (ANZMAC). University of New South Wales, Adelaide.

Lazowski, B., Parker, P., & Rowlands, I. H. (2018). Towards a smart and sustainable residential energy culture: Assessing participant feedback from a long-term smart grid pilot project. *Energy, Sustainability and Society, 8*(1), 1–21.

Manouseli, D., Anderson, B., & Nagarajan, M. (2018). Domestic water demand during droughts in temperate climates: Synthesising evidence for an integrated framework. *Water Resources Management, 32*(2), 433–447.

Mazzocchi, F. (2020). A deeper meaning of sustainability: Insights from indigenous knowledge. *Anthropocene Review, 7*(1), 77–93.

McKague, F., Lawson, R., Scott, M., & Wooliscroft, B. (2016). Understanding the energy consumption choices and coping mechanisms of fuel poor households in New Zealand. *New Zealand Sociology, 31*(1), 106–126.

Minton, E. A., Spielmann, N., Kahle, L. R., & Kim, C. H. (2018). The subjective norms of sustainable consumption: A cross-cultural exploration. *Journal of Business Research, 82*(December 2016), 400–408.

Mirosa, M., Gnoth, D., Lawson, R., & Stephenson, J. (2011). *Rationalizing energy-related behavior in the home: Insights from a value laddering approach.* (June), 2109–2119. European Council for an Energy Efficient Economy Summer Study.

Mirosa, M., Lawson, R., & Gnoth, D. (2013). Linking personal values to energy-efficient behaviors in the home. *Environment and Behavior, 45*(4), 455–475.

Muza, O., & Thomas, V. M. (2022). Cultural norms to support gender equity in energy development: Grounding the productive use agenda in Rwanda. *Energy Research & Social Science, 89*(February), 102543.

Nicholas, L. (2021). *On common ground: How landlords & tenants shape thermal performance of private rental properties.* University of Otago.

Obal, M., Morgan, T., & Joseph, G. (2020). Integrating sustainability into new product development: The role of organizational leadership and culture. *Journal of Small Business Strategy, 30*(1), 43–57.

Oksman, V., Reda, F., Karjalainen, S., & Fatima, Z. (2021). Towards sustainable energy culture in the industrial sector: Introducing an interdisciplinary method for understanding energy culture in business industries. *Energy, Sustainability and Society, 11*(1), 1–13.

Plumwood, V. (2005). *Environmental culture: The ecological crisis of reason.* Routledge.

Rau, H., Moran, P., Manton, R., & Goggins, J. (2020). Changing energy cultures? Household energy use before and after a building energy efficiency retrofit. *Sustainable Cities and Society, 54*, 101983.

Rees, D. (2015). *Modelling transport transitions in New Zealand: Reference guide for model version 1.0.* Centre for Sustainability, University of Otago.

Rotzek, J. N., Scope, C., & Günther, E. (2018). What energy management practice can learn from research on energy culture? *Sustainability Accounting, Management and Policy Journal*, *9*(515–551).

Scott, M. G., & Lawson, R. (2018). The road code: Encouraging more efficient driving practices in New Zealand. *Energy Efficiency, 11*(7), 1617–1626.

Scott, M. G., McCarthy, A., Ford, R., Stephenson, J., & Gorrie, S. (2016). Evaluating the impact of energy interventions: Home audits vs. community events. *Energy Efficiency*, *9*(6), 1221–1240.

Sheth, J. N., & Parvatiyar, A. (2021). Sustainable marketing: Market-driving, not market-driven. *Journal of Micromarketing*, *41*(1), 150–165.

Shove, E., & Spurling, N. (Eds.). (2013). *Sustainable practices: Social theory and climate change*. Routledge.

Snape, J. R., Irvine, K. N., & Rynikiewicz, C. (2011). *Understanding energy behaviours and transitions through the lens of a smart grid Agent Based Model*. Proceedings of the European Council for an Energy Efficient Economy, Summer Study: Energy Efficiency First: The Foundation of a Low-Carbon Society.

Soorige, D., Karunasena, G., Kulatunga, U., Mahmood, M. N., & Silva, L. De. (2022). An energy culture maturity conceptual framework on adopting energy-efficient technology innovations in buildings. *Journal of Open Innovation: Technology, Market, and Complexity*, *8*(60).

Stephenson, J. (2018). Sustainability cultures and energy research: An actor-centred interpretation of cultural theory. *Energy Research & Social Science*, *44*(May), 242–249.

Stephenson, J. (2020). Sustainability Cultures: Exploring the relationships between cultural attributes and sustainability outcomes. In K. Legun, J. Keller, M. Bell, & M. Carolan (Eds.), *Cambridge handchapter of environmental sociology* (pp. 236–248). Cambridge University Press.

Stephenson, J., Hopkins, D., & Doering, A. (2015). Conceptualizing transport transitions: Energy cultures as an organizing framework. *Wiley Interdisciplinary Reviews: Energy and Environment*, *4*(4), 354–364.

Stephenson, J., Lawson, R., Carrington, G., Barton, B., Thorsnes, P., & Mirosa, M. (2010). The practice of interdisciplinarity. *International Journal of Interdisciplinary Social Sciences*, *5*(7), 271–282.

Stephenson, J. R., Sovacool, B. K., & Inderberg, T. H. J. (2021). Energy cultures and national decarbonisation pathways. *Renewable and Sustainable Energy Reviews*, *137*(October).

Stovall, W. R. (2021). *Faces of stone: Environmental history and cultures of Paha Sapa (The Black Hills)*. University of Otago, New Zealand.

Sturgeon, N. L. (2009). *Environmentalism in popular culture: Gender, race, sexuality, and the politics of the natural*. University of Arizona Press.

Sumner, J., Mair, H., & Nelson, E. (2011). Putting the culture back into agriculture: Civic engagement, community and the celebration of local food. *International Journal of Agricultural Sustainability*, *8*(1–2), 54–61.

Tesfamichael, M., Bastille, C., & Leach, M. (2020). Eager to connect, cautious to consume: An integrated view of the drivers and motivations for electricity

consumption among rural households in Kenya. *Energy Research & Social Science, 63*, 101394.
Thorsnes, P., Lawson, R., & Stephenson, J. (2017). Heterogeneity in household preferences for energy-efficient heating systems. *Economics Discussion Papers, 2293*(1713), 40.
Tseng, S. H. Y., Higham, J., & Lee, C. (2022). Academic air travel cultures: A framework for reducing academic flying. In K. Bjørkdahl, A. Santiago, & F. Duharte (Eds.), *Academic flying and the means of communication* (pp. 327–353). Palgrave Macmillan.
Tsing, A. L., Swanson, H. A., Gan, E., & Bubandt, N. (Eds.). (2017). *Arts of living on a damaged planet: Ghosts and monsters of the anthropocene*. University of Minnesota Press.
United Nations. (2015). *Transforming our world: The 2030 agenda for sustainable development*.
van den Berg, L., Roep, D., Hebinck, P., & Teixeira, H. M. (2018). Reassembling nature and culture: Resourceful farming in Araponga, Brazil. *Journal of Rural Studies, 61*, 314–322.
Waldmüller, J. M., Yap, M., & Watene, K. (2022). Remaking the sustainable development goals: relational Indigenous epistemologies. *Policy and Society, 41*(4), 471–485.
Watene, K., & Yap, M. (2015). Culture and sustainable development: Indigenous contributions. *Journal of Global Ethics, 11*(1), 51–55.
Walton, S., Doering, A., Gabriel, C.-A., & Ford, R. (2014). *Energy transitions: Lighting in Vanuatu*. Project report, Australian Aid.
Walton, S., Zhang, A., & O'Kane, C. (2020). Energy eco-innovations for sustainable development: Exploring organizational strategic capabilities through an energy cultures framework. *Business Strategy and the Environment, 29*(3), 812–826.
Welch, D., & Yates, L. (2018). The practices of collective action: Practice theory, sustainability transitions and social change. *Journal for the Theory of Social Behaviour, 48*(3), 288–305.
Yunkaporta, T. (2020). *Sand talk: How indigenous thinking can save the world*. HarperOne.

CHAPTER 9

Conclusion

As I wrote this book, I had four people in mind as my readers: a postgraduate student, a senior policy advisor, the leader of an interdisciplinary research project and an academic in cultural theory. I also wrote it for myself. It is the book I wish I'd had when I was starting out to inquire into what, in the first decade of the twenty-first century, was still mainly referred to as 'behaviour change', certainly in relation to energy research. Behaviour change never sat comfortably with me as a concept—it suggested there was something wrong with people who did not behave in desirable ways, or failed to respond to policy interventions in the ways that had been anticipated. Using culture as a frame of inquiry brings a different lens on how and why people act, think and make decisions, and brings new perspectives on sustainability problems and solutions.

The cultures framework offers a structured approach to cultural analysis. It is designed for the actor-centred exploration of cultural ensembles and their causal relationships with sustainability outcomes. The framework incorporates the concept of agency to account for constraints on actors' capacity to make choices and take action to alter relevant cultural attributes. It draws attention to the ways in which culture is learned and transferred. It accounts for the context in which cultures emerge and are sustained or modified, including the effects of more powerful cultures or structures. The framework invites an orderly exploration of how these variables and their interactions have implications for measures of sustainability.

J. Stephenson, *Culture and Sustainability*,
https://doi.org/10.1007/978-3-031-25515-1_9

I did not anticipate the power and reach of this set of ideas. From a pared-down heuristic initially intended to underpin interdisciplinary research on household energy efficiency, it has proved useful for research on an extraordinary range of topics and is applicable to any actors and any sustainability issue. As previous chapters have elaborated, its applications have produced intrinsic findings about specific cultural ensembles and their outcomes, as well as generic insights into cultural dynamics. The framework helps to explain processes of cultural stasis and cultural change. It has aided studies into gender, power and equity. It is fruitful when used for research design, analysis and integration, and is helpful for the development of policy recommendations. The framework is sufficiently minimalist as to be understandable as a model by non-specialists, while also able to operate at a meta-theoretical level.

Although much of this book has been about the cultures framework, I have also tried to make sense of the many different perspectives on culture. Rather than give preference to a particular interpretation, I have argued that all of these approaches can assist in the momentous and complex transition to a sustainable future. I'd like to finish by sharing some further insights that I have derived while writing this book from examining culture through a sustainability lens and sustainability through a cultural lens.

Sustainability through a Cultural Lens

Unless humans can rapidly transition to more sustainable ways of living with the earth and each other, current and future generations will inherit a world filled with suffering. The biophysical limits of natural systems are already being breached, and climate change, species extinctions, resource depletion, widespread pollution and other consequences are rendering the planet less and less viable for human life. Transitioning requires the dismantling of established institutions and paradigms, global systems of power and financing, expectations of wealth and consumption, chains of debt entanglement and flows of resources and waste. It involves intentionally unpicking long-established norms, beliefs, everyday practices, purchasing decisions, business models, policies, investment choices and material assemblages.

The past transition of equivalent magnitude, the industrial revolution, took much longer and was impelled by the elixir of fossil fuels, new technologies, seemingly boundless resources and the promise of individual wealth creation. This transition is different and much more difficult for three reasons. It requires collective action for the greater good of humanity and for the natural systems we depend on; it will involve an intentional contraction of material expectations for wealthier countries and individuals; and it needs to take place at the very time that planetary destabilisation is starting to have financial and societal consequences. Yet if a deliberate transition to living within planetary limits is not achieved within a few decades, humanity will face a chaotic and even more inequitable future.

In all of this, culture matters. It is the scaffolding of the world's unsustainable trajectory. The interlinked cultural ensembles of colonisation, industrialisation and modernity formed the seemingly unquestionable cultural edifices that have driven the sustainability crisis we face today. As I've discussed in previous chapters, culture is a major barrier to transitioning to a sustainable future because of its tendency to stability. Cultural dynamics can operate to preserve the status quo. Around the world, in all sectors, we can see how more powerful unsustainable cultures can act as structures that force the less powerful to continue harmful patterns of consumption and production.

The only way out of the unsustainable trajectory is through widespread cultural change. What we already know about cultural dynamics suggests that this is not impossible. Cultures *can* change rapidly in response to new circumstances. I am given hope by the many individuals, households, communities, organisations and governments globally that are adopting new ways of thinking, doing and having. These responses are variable in their measurable impact and are based on different perspectives of what sustainability means, but they show how culture can be responsive and innovative in the face of external challenges. I am inspired by traditional and Indigenous cultures whose worldviews encompass respect, reciprocity and relationality with the natural world. I am encouraged to see how new cultural ideas and practices can spread, and the potential for collective cultural change to subvert more powerful cultural ensembles.

To me, the sustainability transition involves cultural change that builds resilience to the consequences of destabilising natural systems and simultaneously reverses the unsustainable outcomes of human endeavours. For many households, that change might include buying fewer or different products, eating lower-carbon foods, making different mobility choices and becoming more responsible for the quality of the local environment. For businesses and organisations, this might mean developing different products and services, becoming part of a circular economy, accounting for carbon flows within the business, changing to different energy sources and paying more attention to worker welfare. But if meta-cultures fail to be transformed as well, these smaller-scale changes will fail to have the impact required. Transition thus also involves challenge and change to culture at multiple levels, including to the powerful actors and their meta-cultures that are still replicating the beliefs and actions that created the sustainability crisis.

Culture is complex, but it is not an excuse for inaction. If it is left in the too-hard basket while the action focuses on technological solutions, sustainability transitions will founder.

Culture through a Sustainability Lens

What does culture comprise, when we are wanting to explore its interplay with sustainability outcomes? I think this depends very much on the question we are wanting to explore, and thetheory of culture we bring to it. As I've previously discussed, there are many perspectives on culture which have originated in different knowledge systems, linguistic applications and academic interpretations. Each of the nine clusters of meaning that I identified in Chapter 2, if applied to a sustainability question, would interpret culture in a different way. Any of these interpretations could contribute to our understanding of how culture operates to enable or constrain sustainability journeys. Rather than being a weakness I see this diversity of perspectives as a latent strength, as discussed at the end of Chapter 8.

When thinking about complex sustainability problems, such as equity, climate change and biodiversity loss, I have come to see the social world as a cacophony of diverse overlapping cultures, from the smallest scale of a household to globally influential structures. Together, these multi-scale cultures are currently largely aligned to defiling and destabilising the world's natural systems, and through that, the future quality of life of human and other species.

From a sustainability perspective, though, it is helpful to remember that culture is mutable. Most people have not freely chosen the cultural ensembles in which they are embedded; culture is learned and, although it may be challenging, people can and do adopt new cultural features. Although cultural actors may feel that the way they live or do business is the only possible way, this is an amnesiac trick of cultural learning. If they jumped back in time or were brought up in a different setting, they could be equally comfortable and satisfied with completely different cultural ensembles. The normalcy of one's own culture is a powerful driver of resistance to change, but it is an illusion.

As I have discussed earlier, there are myriad ways in which cultures can be sustainable. Different places and people will develop distinctive cultures that suit local conditions and local populations. But does a sustainable culture have to be intentionally crafted? Does it matter if cultural actors are consciously aware of the sustainability implications of what they have, think and do? Looking to the past, the best examples of what we might today call sustainable cultures were traditional and Indigenous societies that had learned how to live so as to maintain the health and abundance of natural systems, and had embedded this into knowledge systems, values, spirituality, practices, tools and products. These involve unique concepts that are not precisely the same as what is called sustainability in Western scholarship, but they are great examples of intentional sustainability—where ways of life have been purposefully shaped by an awareness of the need to live with reciprocity and respect for the natural world. Many of these cultures are still extant, although the communities have often been marginalised and their knowledge demeaned by colonisation and modernity.

Then there are those whose lives may be sustainable by some measures, but who are just focused on surviving, living such frugal lives that their environmental impacts are miniscule. They might be thought to be sustainable through some measures, but not in terms of their own health and wellbeing. This reflects the situation of billions of people, mainly in underdeveloped or developing nations, and also the less advantaged in many developed nations. Others, particularly in the first world, may have no option but to live unsustainably, forced by broader influences such as urban form that dictates the need for car dependency. The common factor across all of these is a lack of agency.

Then there are consciously unsustainable cultures. Here I refer to the meta-cultures—the powerful nations, corporations and individuals that are still replicating the systems and processes that created the sustainability crisis and are now pushing our planet's systems to the point of cascading destabilisation. These are not powerless or unaware actors—they are well informed of the implications and yet choose to maintain the status quo in order to benefit their own short-term interests. Even more cynically, many engage in misinformation, obfuscation and inappropriate influence. Consciously unsustainable cultures include actors with significant agency, who know they should change but choose not to do so or use greenwashing to obscure their lack of action. These actors stand in the way of the sustainability transition, normalising complacency and a lack of urgency.

At the same time, new forms of consciously sustainable cultures are emerging, impelled in part by awareness of the gravity of the sustainability crisis and a desire to be part of the solution. These are visible at all scales, from communities establishing shared gardens to businesses adopting circular economies; from the rise in veganism to policies for just transitions; from sustainable fashion to nations committing to rapid zero carbon trajectories. In all instances, these involve the adoption of new ways of thinking, new material choices and new actions and practices.

Ultimately, to achieve a sustainable future, it will be critical to extend the membership, vitality and power of new and established sustainable cultures, diminish those of consciously unsustainable cultures, and support unintentionally unsustainable cultures to extend their agency and develop livelihoods and lifestyles that improve health, wellbeing and equity.

Culture gives me hope. Even though cultural dynamics can tend to replicate unsustainable cultural ensembles, the fact that cultures can and do evolve shows that transformational journeys are possible. Cultural variability shows that people can live simply, sustainably and happily in a wide range of environmental conditions. It shows how a multiplicity of cultural arrangements, tailored to place and people, will be needed for a sustainable future. Culture gives me hope because cultural actors can rapidly and creatively adjust their ensembles to respond to new circumstances or in anticipation of shocks. I also gain hope from seeing the potential of collective cultural change to destabilise more powerful cultures and structures that reinforce the status quo. The sustainability crisis requires that these cultural processes are better understood and become more visible.

Culture and Sustainability

Culture is the missing link when it comes to the sustainability transition. There are vast reservoirs of knowledge about technical solutions, policy options, economic tools and human psychology, but relatively little is known about the role of culture. Despite its fundamental influence on unsustainable systems of production and consumption, culture is often overlooked, ignored or glossed over. Its apparent intangibility and complexity become a reason for inaction. Yet culture does not need to be a mystery.

Cultural analysis can provide insights that are not achievable through other lenses, reveal possibilities for change and give impetus to efforts for a sustainable future. It can reveal processes of social stability (resistance to change), social fluidity (responding to a changing context) and social reorientation (purposeful transformation). We can get trapped into thinking the human race is doomed because of problematic psychological tendencies such as an inability to consider the needs of future generations, forgetting that plenty of cultures already embed intergenerational thinking. Cultural analysis can help augment the positive role of culture in sustainability transitions: solving problems by dynamically adjusting to changing external circumstances; generating and rapidly proliferating new ideas, practices, knowledge and products; being a source of traditional knowledge, values and practices that are aligned with a sustainable future; and delivering collectively desired outcomes.

The cultures framework works at this nexus of culture and sustainability. As a stripped-down model of culture's relationship with sustainability, the framework is for anyone to use. It can help individuals and households to notice their own cultural ensembles. It can help businesses and organisations look reflexively at themselves and see where to start a journey of purposeful transformation. It can help governance institutions and policymakers to craft more holistic policies and interventions, and to value cultural actors who are already demonstrating what it means to live sustainably. More generally, it helps explain why and how people matter in conversations that are usually dominated by technology and economics.

For my postgraduate student reader, I offer this book as a resource to help you design and undertake your research. It brings the cultures framework up to date, drawing inspiration from the many researchers around the world globally who have used it as well as from its diverse applications by the original Energy Cultures team and my own postgraduate students. My talks to students and researchers on using the framework to undertake research has informed the step-by-step guide in Chapter 8. I have tried to cover all the questions that my own students have asked, and queries that have come in from other postgraduate students around the world. You can find further guidance in the various studies I have cited and the theories and methodologies they have used. I hope you are inspired to explore new questions of culture and sustainability, and I look forward to seeing your work when it is published.

For my policy-oriented reader, I hope that the framework helps you to see how and why behaviour-change policies are often not as successful as you would hope. Taking a cultural perspective helps to explain those thorny issues for policy such as value-action gaps, rebound effects and non-rational responses to financial signals. It helps you to see why some policies might have unintended consequences, and why other policies might face stiff resistance from your intended audience. Once you understand the entanglements between what people think, do and have, it becomes clear why an intervention to change one of these can have consequences for the others, or why its effectiveness might be constrained. I hope that the step-by-step guide in Chapter 7 can help you to develop culturally informed policies that account for the heterogeneity of cultural ensembles relevant to the problem you are seeking to address. Accounting for cultural dynamics, agency, cultural vectors and external influences can help you to design targeted policies that respond better to the complexities of everyday existence for households and organisations. Once you have designed your policy, the framework also offers a scaffold around which to develop evaluations of the effectiveness of interventions. I trust that this book has also opened your eyes to the importance of supporting and learning from cultures that are already sustainable or are on self-initiated journeys of sustainability.

For my would-be leader of a research programme: you already see the critical importance of interdisciplinary research to address complex sustainability problems, but you may be wondering how to develop a programme of research that benefits from and integrates the varied contributions of your team members. The cultures framework can be useful as a core integrative structure for your research. Your team can use the framework to help depict the research problem; develop hypotheses regarding the relationship between cultural ensembles and outcomes; identify the key questions and variables to study; specify team members' roles and the theories and methodologies they will use to explore variables and their dynamics; triangulate and integrate findings; and identify opportunities for supporting positive change. With your large programme you may have capacity to explore heterogeneity across cultures and the nature and influence of multi-level cultures. I hope that in respecting the contributions of many disciplines, you will extend that respect to knowledge-holders from outside of traditional Western disciplines. Local and Indigenous knowledge has a huge amount to offer to the pursuit of sustainability.

For my academic in cultural theory, I offer this book with some trepidation. Each of the fields of cultural scholarship that I discuss in Chapters 2 and 3 would require a lifetime's worth of study to fully comprehend, and there is no way that I can do it justice. I may offend some of your colleagues by my seemingly scanty representations of rich bodies of work and their many shades of meaning and contention. I hope that you can see that this is because my intention was to be integrative rather than reductive, spanning rather than deep-diving. I sought to draw out high-level differences and similarities, and explore the potential for all forms of cultural theorising to contribute to questions of sustainability. My broad categorisations of theories of culture, including Indigenous theories, may offer some novel perspectives for cultural scholars. What excites me most, and I hope you as well, is the potential for all cultural theories to make a much greater contribution to resolving global and local sustainability challenges.

To all of you, and for other readers who have dipped into this book from personal interest or concern about the world's unsustainable trajectory, I hope you can now 'see' the operations of culture in the sustainability outcomes of your own lives or of the organisations for whom you work. Understanding the multiple overlapping cultures of your own consumption and production, and the external influences that shape them, can reveal why it is so hard to extricate yourself from unsustainable patterns. I hope it alerts you to the consequences of norms, beliefs, meanings, forms of knowledge and other shared aspects of cognition, and how these inexorably shape what you do and what you acquire and make. I hope it is easier to see how powerful but largely invisible cultures have quietly shaped your own culture and the world we have inherited. I hope it helps you to unpick and analyse the workings of culture at multiple levels and to see that it is possible to take collective action to reach beyond the agency barrier and re-shape wider unsustainable structures. There is huge promise in getting to know ourselves as cultural creatures and understanding the sustainability transition as a cultural transition.

Index

J. Stephenson, *Culture and Sustainability*,
https://doi.org/10.1007/978-3-031-25515-1

D

E

The manufacturer's authorised representative in the EU is Springer Nature Customer Service Centre GmbH, Europaplatz 3, 69115 Heidelberg, Germany. If you have any concerns regarding our products, please contact ProductSafety@springernature.com

Printed and bound by CPI Group (UK) Ltd, Croydon, CR0 4YY

28/09/2026

02224072-0005